METHODS IN MOLECULAR BIOLOGY

Series Editor
John M. Walker
School of Life Sciences
University of Hertfordshire
Hatfield, Hertfordshire, AL10 9AB, UK

For further volumes:
http://www.springer.com/series/7651

Drug Delivery System

Second Edition

Edited by

Kewal K. Jain

Jain PharmaBiotech, Basel, Switzerland

 Humana Press

Editor
Kewal K. Jain
Jain PharmaBiotech
Basel, Switzerland

1007168221

ISSN 1064-3745   ISSN 1940-6029 (electronic)
ISBN 978-1-4939-0362-7   ISBN 978-1-4939-0363-4 (eBook)
DOI 10.1007/978-1-4939-0363-4
Springer New York Heidelberg Dordrecht London

Library of Congress Control Number: 2014931849

Printed on acid-free paper

Humana Press is a brand of Springer
Springer is part of Springer Science+Business Media (www.springer.com)

Preface

Several new technologies for drug delivery have been developed since the publication of the first edition of this book 5 years ago. Some of these are described in the second edition in the style that has led to the popularity of "Methods in Molecular Biology" series. The overview chapter has been updated to include references to new methods including those that are described in this book. Some of the novel methods of drug delivery are published here for the first time. There is an increase in the refinement and use of nanobiotechnology techniques for drug delivery, which form the basis of 6 of the 15 techniques described. Most of the methods are experimental and used in laboratories, but one method in clinical use for the intrathecal delivery of analgesics is described in detail. It is hoped that this book will continue to be useful for pharmaceutical scientists as well as physicians both in the academic institutions and in the industry.

Basel, Switzerland *Kewal K. Jain, M.D.*

Contents

Contributors

D. AGUDELO • *Department of Chemistry–Physics, University of Québec at Trois-Rivières, Trois-Rivières, QC, Canada*

AHLAM ZAID ALKILANI • *Medical Biology Centre, School of Pharmacy, Queen's University Belfast, Belfast, UK*

JUAN MARCOS ASENSIO-SAMPER • *Multidisciplinary Pain Management Department, Valencia University General Hospital, Valencia, Spain; Department of Anesthesia, Valencia University General Hospital, Valencia, Spain*

SURINDER BATRA • *Department of Biochemistry and Molecular Biology, College of Medicine, University of Nebraska Medical Center, Omaha, NE, USA; Eppley Institute for Research in Cancer and Allied Diseases, University of Nebraska Medical Center, Omaha, NE, USA; Buffet Cancer Center, Omaha, NE, USA*

MORITZ BECK-BROICHSITTER • *Faculté de Pharmacie, Institut Galien, Université Paris-Sud, Châtenay-Malabry, France; Department of Internal Medicine, Medical Clinic II, Justus-Liebig-Universität, Giessen, Germany*

NATHAN M. BELLIVEAU • *Precision NanoSystems Inc., Vancouver, BC, Canada*

ARCHANA V. BOOPATHY • *Wallace H. Coulter Department of Biomedical Engineering, Emory University and Georgia Institute of Technology, Atlanta, GA, USA*

JASON CASTLE • *GE Global Research Niskayuna, Schenectady, NY, USA*

P. CHANPHAI • *Department of Chemistry–Physics, University of Québec at Trois-Rivières, Trois-Rivières, QC, Canada*

SAM CHEN • *University of British Columbia, Vancouver, Vancouver, BC, Canada*

MICHELLE CRONIN • *Cork Cancer Research Centre, BioSciences Institute, University College Cork, Cork, Ireland*

PIETER R. CULLIS • *University of British Columbia, Vancouver, Vancouver, BC, Canada; Precision NanoSystems Inc., Vancouver, Vancouver, BC, Canada*

JOANNE CUMMINS • *Cork Cancer Research Centre, BioSciences Institute, University College Cork, Cork, Ireland*

ALEXANDRA C. DALLA-BONA • *Department of Internal Medicine, Medical Clinic II, Justus-Liebig-Universität, Giessen, Germany*

MICHAEL E. DAVIS • *Wallace H. Coulter Department of Biomedical Engineering, Emory University and Georgia Institute of Technology, Atlanta, GA, USA*

ARNAB DE • *Department of Microbiology and Immunology, Columbia University, New York, NY, USA*

JOSE DE ANDRES • *Multidisciplinary Pain Management Department, Valencia University General Hospital, Valencia, Spain; Department of Anesthesia, Valencia University General Hospital, Valencia, Spain*

RYAN F. DONNELLY • *Medical Biology Centre, School of Pharmacy, Queen's University Belfast, Belfast, UK*

GUSTAVO FABREGAT-CID • *Multidisciplinary Pain Management Department, Valencia University General Hospital, Valencia, Spain; Department of Anesthesia, Valencia University General Hospital, Valencia, Spain*

STEVEN B. FEINSTEIN • *Rush University medical Center, Chicago, IL, USA*

CORMAC G.M. GAHAN • *Department of Microbiology and Alimentary Pharmabiotic Centre, University College Cork, Cork, Ireland; School of Pharmacy, University College Cork, Cork, Ireland*

JIAN-QING GAO • *Institute of Pharmaceutics, College of Pharmaceutical Sciences, Zhejiang University, Hangzhou, Zhejiang, P.R. China*

SEEMA GARG • *Department of Chemistry, University of Delhi, Delhi, India*

MARTIN J. GARLAND • *Medical Biology Centre, School of Pharmacy, Queen's University Belfast, Belfast, UK*

DANIELA GUARNIERI • *Center for Advanced Biomaterials for Health Care@CRIB, Italian Institute of Technology, IIT, Naples, Italy*

SUPRIT GUPTA • *Department of Biochemistry and Molecular Biology, College of Medicine, University of Nebraska Medical Center, Omaha, NE, USA*

CARL L. HANSEN • *University of British Columbia, Vancouver, Vancouver, BC, Canada; Precision NanoSystems Inc., Vancouver, Vancouver, BC, Canada*

HIDEYOSHI HARASHIMA • *Laboratory for Molecular Design of Pharmaceutics, Faculty of Pharmaceutical Sciences, Hokkaido University, Sapporo, Japan*

YU-LAN HU • *Institute of Pharmaceutics, College of Pharmaceutical Sciences, Zhejiang University, Hangzhou, Zhejiang, P.R. China*

JENS HUFT • *University of British Columbia, Vancouver, BC, Canada*

KEWAL K. JAIN • *Jain PharmaBiotech, Basel, Switzerland*

MANEESH JAIN • *Department of Biochemistry and Molecular Biology, College of Medicine, University of Nebraska Medical Center, Omaha, NE, USA*

THOMAS KISSEL • *Department of Pharmaceutics and Biopharmacy, Philipps-Universität, Marburg, Germany*

TIM J. LEAVER • *Precision NanoSystems Inc., Vancouver, BC, Canada*

JUSTIN B. LEE • *University of British Columbia, Vancouver, Vancouver, BC, Canada*

ALEX K. LEUNG • *University of British Columbia, Vancouver, Vancouver, BC, Canada*

PAULO J. LIN • *University of British Columbia, Vancouver, Vancouver, BC, Canada*

MEENAKSHI MALHOTRA • *Pharmacodelivery Group, School of Pharmacy, University College Cork, Cork, Ireland; Biomedical Technology and Cell Therapy Research Laboratory, Departments of Biomedical Engineering, Faculty of Medicine, McGill University, Montreal, QC, Canada*

SUSHIL MISHRA • *Department of Chemistry, University of Delhi, Delhi, India*

SUBHO MOZUMDAR • *Department of Chemistry, University of Delhi, Delhi, India*

ORNELLA MUSCETTI • *Center for Advanced Biomaterials for Health Care@CRIB and Interdisciplinary Research Centre on Biomaterials (CRIB), Italian Institute of Technology, IIT and University of Naples Federico II, Naples, Italy*

SH. NAFISI • *Department of Chemistry, San Jose State University, San Jose, CA, USA*

PAOLO A. NETTI • *Center for Advanced Biomaterials for Health Care@CRIB and Interdisciplinary Research Centre on Biomaterials (CRIB), Italian Institute of Technology, IIT and University of Naples Federico II, Naples, Italy*

KEVIN OU • *University of British Columbia, Vancouver, BC, Canada*

SATYA PRAKASH • *Biomedical Technology and Cell Therapy Research Laboratory, Department of Biomedical Engineering, Faculty of Medicine, McGill University, Montreal, QC, Canada*

EUAN C. RAMSAY • *University of British Columbia, Vancouver, Vancouver, BC, Canada; Precision NanoSystems Inc., Vancouver, Vancouver, BC, Canada*

SHYAMALI SAHA • *Biomedical Technology and Cell Therapy Research Laboratory, Department of Biomedical Engineering, Faculty of Medicine, McGill University, Montreal, QC, Canada; Faculty of Dentistry, McGill University, Montreal, QC, Canada*

S. SANYAKAMDHORN • *Department of Chemistry–Physics, University of Québec at Trois-Rivières, Trois-Rivières, QC, Canada*

THOMAS SCHMEHL • *Department of Internal Medicine, Medical Clinic II, Justus-Liebig-Universität, Giessen, Germany*

WERNER SEEGER • *Department of Internal Medicine, Medical Clinic II, Justus-Liebig-Universität, Giessen, Germany*

H.A. TAJMIR-RIAHI • *Department of Chemistry–Physics, University of Québec at Trois-Rivières, Trois-Rivières, QC, Canada*

MARK TANGNEY • *Cancer Research Centre, BioSciences Institute, University College Cork, Cork, Ireland*

ROBERT J. TAYLOR • *Precision NanoSystems Inc., Vancouver, Vancouver, BC, Canada*

CATHERINE TOMARO-DUCHESNEAU • *Biomedical Technology and Cell Therapy Research Laboratory, Department of Biomedical Engineering, Faculty of Medicine, McGill University, Montreal, QC, Canada*

JAN PETER VAN PIJKEREN • *Department of Microbiology and Molecular Genetics, Michigan State University, East Lansing, MI, USA*

COLIN WALSH • *University of British Columbia, Vancouver, BC, Canada*

ANDRE W. WILD • *University of British Columbia, Vancouver, BC, Canada*

YUMA YAMADA • *Laboratory for Molecular Design of Pharmaceutics, Faculty of Pharmaceutical Sciences, Hokkaido University, Sapporo, Japan*

Chapter 1

Current Status and Future Prospects of Drug Delivery Systems

Kewal K. Jain

Abstract

This is an overview of the current drug delivery systems (DDS) starting with various routes of drug administration. Various drug formulations are then described as well as devices used for drug delivery and targeted drug delivery. There has been a considerable increase in the number of new biotechnology-based therapeutics. Most of these are proteins and peptides, and their delivery presents special challenges. Cell and gene therapies are sophisticated methods of delivery of therapeutics. Nanoparticles are considered to be important in refining drug delivery; they can be pharmaceuticals as well as diagnostics. Refinements in drug delivery will facilitate the development of personalized medicine in which targeted drug delivery will play an important role. There is discussion about the ideal DDS, commercial aspects, challenges, and future prospects.

Key words Cell therapy, Controlled release, Drug delivery systems, Drug formulations, Gene therapy, Nanobiotechnology, Nanoparticles, Personalized medicine, Protein/peptide delivery, Routes of drug administration, Targeted drug delivery

1 Introduction

A drug delivery system (DDS) is defined as a formulation or a device that enables the introduction of a therapeutic substance in the body and improves its efficacy and safety by controlling the rate, time, and place of release of drugs in the body. This process includes the administration of the therapeutic product, the release of the active ingredients by the product, and the subsequent transport of the active ingredients across the biological membranes to the site of action. The term therapeutic substance also applies to an agent such as gene therapy that will induce in vivo production of the active therapeutic agent. Gene therapy can fit in the basic and broad definition of a drug delivery system. Gene vectors may need to be introduced into the human body by novel delivery methods. However, gene therapy has its own special regulatory control.

Kewal K. Jain (ed.), *Drug Delivery System*, Methods in Molecular Biology, vol. 1141,
DOI 10.1007/978-1-4939-0363-4_1, © Springer Science+Business Media New York 2014

Drug delivery system is an interface between the patient and the drug. It may be a formulation of the drug to administer it for a therapeutic purpose or a device used to deliver the drug. This distinction between the drug and the device is important as it is the criterion for regulatory control of the delivery system by the drug or medicine control agency. If a device is introduced into the human body for purpose other than drug administration, such as therapeutic effect by a physical modality or a drug may be incorporated into the device for preventing complications resulting from the device, it is regulated strictly as a device. There is a wide spectrum between the drugs and devices, and the allocation to one or the other category is decided on a case-by-case basis.

2 Drug Delivery Routes

Drug may be introduced into the human body by various anatomical routes. They may be intended for systemic effects or targeted to various organs and diseases. The choice of the route of administration depends on the disease, the effect desired, and the product available. Drugs may be administered directly to the organ affected by disease or given systemically and target to the diseased organ. A classification of various methods of systemic drug delivery by anatomical routes is shown in Table 1.

Table 1
A classification of various anatomical routes for systemic drug delivery

Gastrointestinal system

  Oral

  Rectal

Parenteral

  Subcutaneous injection

  Intramuscular injection

  Intravenous injection

  Intra-arterial injection

Transmucosal: buccal and through mucosa lining the rest of gastrointestinal tract

Transnasal

Pulmonary: drug delivery by inhalation

Transdermal drug delivery

Intra-osseous infusion

2.1   Oral Drug Delivery

Historically, the oral route of drug administration has been the one used most for both conventional and novel drug delivery. The reasons for this preference are obvious: ease of administration and widespread acceptance by the patients. Major limitations of oral route of drug administration are:

- Drugs taken orally for systemic effects have variable absorption rate and variable serum concentrations which may be unpredictable. This has led to the development of sustained-release and controlled-release systems.

- The high acid content and ubiquitous digestive enzymes of the digestive tract can degrade some drugs well before they reach the site of absorption into the bloodstream. This is a particular problem for ingested proteins. Therefore, this route has limitations for the administration of biotechnology products.

- Many macromolecules and polar compounds cannot effectively traverse the cells of the epithelial membrane in the small intestines to reach the bloodstream. Their use is limited to local effect in the gastrointestinal tract.

- Many drugs become insoluble at the low pH levels encountered in the digestive tract. Since only the soluble form of the drug can be absorbed into the bloodstream, the transition of the drug to the insoluble form can significantly reduce bioavailability.

- The drug may be inactivated in the liver on its way to the systemic circulation. An example of this is the inactivation of glyceryl trinitrate by hepatic monooxygenase enzymes during the first-pass metabolism.

- Some drugs irritate the gastrointestinal tract, and this is partially counteracted by coating.

- Oral route may not be suitable for drugs targeted to specific organs.

- Despite disadvantages, the oral route remains the preferred route of drug delivery. Several improvements have taken place in the formulation of drugs for oral delivery for improving the action.

2.2   Parenteral Drug Delivery

Parenteral literally means introduction of substances into the body by routes other than the gastrointestinal tract, but practically the term is applied to injection of substances by subcutaneous, intramuscular, intravenous, and intra-arterial routes. Injections made into specific organs of the body for targeted drug delivery will be described under various therapeutic areas.

Parenteral administration of the drugs is now an established part of medical practice and is the most commonly used invasive method of drug delivery. Many important drugs are available only in parenteral form. Conventional syringes with needles are either glass or plastic (disposable). Nonreusable syringe and needle come either with autodestruct syringes which lock after injection or retractable needles. Advantages of parenteral administration are:

- Rapid onset of action
- Predictable and almost complete bioavailability
- Avoidance of the gastrointestinal tract with problems of oral drug administration
- Provides a reliable route for drug administration in very ill and comatose patients, who are not able to ingest anything orally

Major drawbacks of parenteral administration are:

- Injection is not an ideal method of delivery because of pain involved and patient compliance becomes a major problem.
- Injections have limitations for the delivery of protein products particularly those that require sustained levels.

Comments on various types of injections are given in the following text.

Subcutaneous. This involves the introduction of the drug to a layer to subcutaneous fatty tissue by the use of a hypodermic needle. Large portions of the body are available for subcutaneous injection which can be given by the patients themselves as in the case of insulin for diabetes. Various factors which influence drug delivery by subcutaneous route are:

- Size of the molecules as the larger molecules have slower penetration rates than smaller ones.
- Viscosity may impede the diffusion of drugs into body fluids.
- The anatomical characteristics of the site of injection such as vascularity and amount of fatty tissue influence the rate of absorption of the drug.

Subcutaneous injections usually have a lower rate of absorption and slower onset of action than intramuscular or intravenous injections. The rate of absorption may be enhanced by infiltration with the enzyme hyaluronidase. Disadvantages of subcutaneous injection are:

- The rate of absorption is difficult to control from the subcutaneous deposit.
- Local complications which include irritation and pain at site of injection.
- Injection sites have to be changed frequently to avoid accumulation of the unabsorbed drug which may cause tissue damage.

Several self-administration subcutaneous injection systems are available and include conventional syringes, pre-filled glass syringes, autoinjectors, pen pumps and needle-less injectors. Subcutaneous still remains predictable and controllable route of delivery for peptides and macromolecules.

Intramuscular Injections. These are given deep into skeletal muscles, usually the deltoids or the gluteal muscles. The onset of action after intramuscular injection is faster than with subcutaneous injection but slower than with intravenous injection. The absorption of the drug is diffusion controlled, but it is faster due to high vascularity of the muscle tissue. The rate of absorption varies according to physico-chemical properties of the solution-injected and physiological variables such as blood circulation of the muscle and the state of muscular activity. Disadvantages of intramuscular route for drug delivery are:

- Pain at the injection site.
- Limitation of the amount injected according to the mass of the muscle available.
- Degradation of peptides at the site of injection.
- Complications include peripheral nerve injury and formation of hematoma and abscess at the site of injection.
- Inadvertent puncture of a blood vessel during injection may introduce the drug directly into the blood circulation.

Most injectable products can be given intramuscularly. Numerous dosage forms are available for this route: oil-in-water emulsions, colloidal suspensions, and reconstituted powders. The product form in which the drug is not fully dissolved generally results in slower, more gradual absorption and slower onset of action with longer-lasting effects. Intramuscularly administered drugs typically form a depot in the muscle mass from which the drug is slowly absorbed. Peak drug concentrations are usually seen from 1 to 2 h. Factors which affect the rate of release of the drug from such a depot include the following:

- Compactness of the depot as the release is faster from a less compact and more diffuse depot
- Concentration and particle size of drug in the vehicle
- Nature of solvent in the injection
- Physical form of the product
- The flow characteristics of the product
- Volume of the injection

Intravenous Administration. This involves injection in the form of an aqueous into a superficial vein or continuous infusion via a needle or a catheter placed in a superficial or deep vein. This is the only method of administration available for some drugs and is chosen in emergency situations because the onset of action is rapid following the injection. Theoretically, none of the drug is lost; smaller doses are required than with other routes of administration. The rate of infusion can be controlled for prolonged and continuous administration. Devices are available for timed administration of intermittent doses via an intravenous catheter. The particles in the intravenous solution are distributed to various organs depending on the particle size. Particles larger than 7 μm are trapped in the lungs, and those smaller than 0.1 μm accumulate in the bone marrow. Those with diameter between 0.1 and 7 μm are taken up by the liver and the spleen. This information is useful in targeting of a drug to various organs. Disadvantages of the intravenous route are:

- Immune reactions may occur following injections of proteins and peptides.
- Trauma to veins can lead to thrombophlebitis.
- Extravasation of the drug solution into the extravascular space may lead to irritation and tissue necrosis.
- Infections may occur at the site of catheter introduction.
- Air embolism may occur due to air sucked in via the intravenous line.

It is now possible to modify the kinetics of disposition and sometimes the metabolic profile of a drug given by intravenous route. This can be achieved by incorporating the drug into nanovesicles such as liposomes.

Intra-arterial. Direct injection into the arteries is not a usual route for therapeutic drug administration. Arterial puncture and injection of contrast material has been carried out for angiography. Most of the intra-arterial injections or arterial perfusions via catheters placed in arteries are for regional chemotherapy of some organs and limbs. Intra-arterial chemotherapy has been used for malignant tumors of the brain.

2.3 Transdermal Drug Delivery

Transdermal drug delivery is an approach used to deliver drugs through the skin for therapeutic use as an alternative to oral, intravascular, subcutaneous, and transmucosal routes. A detailed description of technologies and commercial aspects of development are described in a special report on this topic [1].

It includes the following categories of drug administration:

- Local application formulations, e.g., transdermal gels
- Penetration enhancers
- Drug carriers, e.g., liposomes and nanoparticles
- Transdermal patches
- Transdermal electrotransport

- Use of physical modalities to facilitate transdermal drug transport
- Minimally invasive methods of transdermal drug delivery, e.g., needle-free injections

Iontophoresis and microneedles are playing an increasing role in transdermal drug delivery. A technique has been described using hydrogel-forming microneedle arrays in combination with electrophoresis for controlled transdermal delivery of biomacromolecules in a simple, one-step approach [2]. It broadens the range of drugs administered transdermally.

2.4 Transmucosal Drug Delivery

Mucous membrane covers all the internal passages and orifices of the body, and drugs can be introduced at various anatomical sites. Only some general statements applicable to all mucous membranes will be made here, and the details will be described according to the locations such as buccal, nasal, rectal, etc.

Movement of penetrants across the mucous membranes is by diffusion. At steady state, the amount of a substance crossing the tissue per unit of time is constant, and the permeability coefficients are not influenced by the concentration of the solutions or the direction of nonelectrolyte transfer. As in case of the epidermis of the skin, the pathways of permeation through the epithelial barriers are intercellular rather than intracellular. The permeability can be enhanced by the use of surfactants such as sodium lauryl sulfate (a cationic surfactant). An unsaturated fatty acid, oleic acid, in a propylene glycol vehicle can act as a penetration enhancer for diffusion of propranolol through the porcine buccal mucosa in vitro. Delivery of biopharmaceuticals across mucosal surfaces may offer several advantages over injection techniques which include the following:

- Avoidance of an injection
- Increase of therapeutic efficiency
- Possibility of administering peptides
- Rapid absorption as compare to oral administration
- Bypassing first-pass metabolism by the liver
- Higher patient acceptance as compared to injectables
- Lower cost than injections

Mucoadhesive controlled-release devices can improve the effectiveness of transmucosal delivery of a drug by maintaining the drug concentration between the effective and toxic levels, inhibiting the dilution of the drug in the body fluids, and allowing targeting and localization of a drug at a specific site. Acrylic-based hydrogels have been used extensively as mucoadhesive systems. They are well suited for bioadhesion due to their flexibility and nonabrasive characteristics in the partially swollen state, which reduce damage-causing attrition to the tissues in contact. Cross-linked polymeric devices may be rendered adhesive to the mucosa. For example, adhesive capabilities of these hydrogels can be improved by tethering of long flexible poly(ethylene glycol) chains. The ensuing hydrogels exhibit mucoadhesive properties due to enhanced anchoring of the chains with the mucosa.

Buccal and Sublingual Routes. Buccal absorption is dependent on lipid solubility of the non-ionized drug, the salivary pH, and the partition coefficient which is an index of the relative affinity of the drug for the vehicle compared to the epithelial barrier. A large partition coefficient value indicates a poor affinity the vehicle for the drug. A small partition coefficient value means a strong interaction between the drug and the vehicle which reduces the release of the drug from the vehicle. The ideal vehicle is the one in which the drug is minimally soluble. Buccal drug administration has the following attractive features:

- Quick absorption into the systemic circulation with rapid onset of effect due to absorption from the rich mucosal network of systemic veins and lymphatics.
- The tablet can be removed in case of an undesirable effect.
- Oral mucosal absorption avoids the first-pass hepatic metabolism.
- A tablet can remain for a prolonged period in the buccal cavity which enables development of formulations with sustained-release effect.
- This route can be used in patients with swallowing difficulties.

  Limitations to the use of buccal route are:

- The tablet must be kept in place and not chewed or swallowed.
- Excessive salivary flow may cause too rapid dissolution and absorption of the tablet or wash it away.
- A bad-tasting tablet will have a low patient acceptability.
- Some of these disadvantages have been overcome by the use of a patch containing the drug which is applied to the buccal mucosa or the using the drug as a spray.

2.5   Nasal Drug Delivery

Drugs have been administered nasally for several years for both topical and systemic effect. Topical administration includes agents for the treatment of nasal congestion, rhinitis, sinusitis, and related allergic and other chronic conditions. Various medications include corticosteroids, antihistamines, anticholinergics, and vasoconstrictors. The nasal route is an attractive target for administration of the drug of choice, particularly in overcoming disadvantages such as high first-pass metabolism and drug degradation in the gastrointestinal tract associated with the oral administration. The focus in recent years has been on the use of nasal route for systemic drug delivery, but it can also be used for direct drug delivery to the brain.

Surface Epithelium of the Nasal Cavity. The anterior one-third of the nasal cavity is covered by a squamous and transitional epithelium, the upper part of the cavity by an olfactory epithelium and the remaining portion by a typical airway epithelium, which is ciliated, pseudostratified, and columnar. The columnar cells are related to neighboring cells by tight junctions at the apices as well as by interdigitations of the cell membrane. The cilia have an important function for propelling the mucous into the throat. Toxic effect of the drug on the cilia impairs the mucous clearance. Safety of drugs for nasal delivery has been studied by in vitro effect on ciliary beating and its reversibility as well as on the physical properties of the mucous layer.

Intranasal Drug Delivery. Intranasal route is considered for drugs which are ineffective orally, are used chronically, and require small doses, and rapid entry into the circulation is desired. The rate of diffusion of the compounds through the nasal mucous membranes, like other biological membranes, is influenced by the physico-chemical properties of the compound. However, in vivo nasal absorption of compounds of molecular weight less than 300 is not significantly influenced by the physicochemical properties of the drug. Factors such as the size of the molecule and the ability of the compound to hydrogen bond with the component of the membrane are more important than lipophilicity and ionization state. The absorption of drugs from the nasal mucosa most probably takes place via the aqueous channels portion of the membrane. Therefore, as long as the drug is in solution and the molecular size is small, the drug will be absorbed rapidly via the aqueous path of the membrane. The absorption from the nasal cavity decreases as the molecular size increases. Factors which affect the rate and extent of absorption of drugs via the nasal route are:

- The rate of nasal secretion. The greater the rate of secretion, the smaller the bioavailability of the drug.

- Ciliary movement. The faster the ciliary movement, the smaller the bioavailability of the drug.

- Vascularity of the nose. Increase of blood flow leads to faster drug absorption and vice versa.

- Metabolism of drugs in the nasal cavity. Although enzymes are found in the nasal tissues, they do not significantly affect the absorption of most compounds except peptides which can be degraded by aminopeptidases. This may be due to low levels of enzymes and short exposure time of the drug to the enzyme.

- Diseases affecting nasal mucous membrane. Effect of the common cold on nasal drug absorption is also an important consideration.
- A major limitation of nasal delivery is rapid mucociliary clearance resulting in low absorption and hence poor bioavailability of the drug. In situ nasal gelling drug delivery systems have been explored to overcome this and provide sustained delivery via nasal route [3].

2.5.1 Enhancement of Nasal Drug Delivery

Nasal drug delivery can be enhanced by reducing drug metabolism, by prolonging the drug residence time in the nasal cavity, and by increasing absorption. The last is the most important strategy and will be discussed here.

Nasal drug absorption can be accomplished by use of prodrugs, chemical modification of the parent molecule, and use of physical methods of increasing permeability. Special excipient used in the nasal preparations come into contact with the nasal mucosa and may exert some effect to facilitate the drug transport. The mucosal pores are easier to open than those in the epidermis. The following characteristics should be considered in choosing an absorption enhancer:

- The enhancer should be pharmacologically inert.
- It should be non-irritating, nontoxic, and nonallergic.
- Its effect on the nasal mucosa should be reversible.
- It should be compatible with the drug.
- It should be able to remain in contact with the nasal mucosa long enough to achieve maximal effects.
- It should not have any offensive odor or taste.
- It should be relatively inexpensive and readily available.

The effect of nasal absorption enhancers on ciliary beating needs to be tested as any adverse effect on mucociliary clearance will limit the patient's acceptance of the nasal formulation. Chitosan, a naturally occurring polysaccharide that is extracted from the shells of crustaceans, is an absorption enhancer. Chitosan is bioadhesive and binds to the mucosal membrane, prolonging the retention time of the formulation on the nasal mucosa. Chitosan may also facilitate absorption through promoting paracellular transport or through other mechanisms. The chitosan nasal technology can be exploited as solution, dry powders, or microsphere formulations to further optimize the delivery system for individual compounds. Impressive improvements in bioavailability have been achieved with a range of compounds. For compounds requiring rapid onset of action, the nasal chitosan technology can provide a fast peak concentration compared with oral or subcutaneous administration.

2.5.2 Advantages of Nasal Drug Delivery

- High permeability of the nasal mucosa compared to the epidermis or the gastrointestinal mucosa
- Highly vascularized subepithelial tissue
- Rapid absorption, usually within half an hour
- Avoidance of first-pass effect that occurs after absorption of drugs from the gastrointestinal tract
- Avoidance of the effects of gastric stasis and vomiting, for example, in migraine patients
- Ease of administration by the patients who are usually familiar with nasal drops and sprays
- Higher bioavailability of the drugs than in case of gastrointestinal route or pulmonary route
- Most feasible route for the delivery of peptides

2.5.3 Disadvantages of Nasal Drug Delivery

- Diseases conditions of the nose may result in impaired absorption.
- Dose is limited due to relatively small area available for absorption.
- Time available for absorption is limited.
- Little is known of the effect of common cold on transnasal drug delivery, and it is likely that instilling a drug into a blocked nose or a nose with surplus of watery rhinorrhea may expel the medication from the nose.
- The nasal route of delivery is not applicable to all drugs. Polar drugs and some macromolecules are not absorbed in sufficient concentration due to poor membrane permeability, rapid clearance, and enzymatic degradation into the nasal cavity.

Alternative means that help overcome these nasal barriers are currently in development. Absorption enhancers such as phospholipids and surfactants are constantly used, but care must be taken in relation to their concentration. Drug delivery systems including liposomes, cyclodextrins, and micro- and nanoparticles are being investigated to increase the bioavailability of drugs delivered intranasally.

After a consideration of advantages as well disadvantages, nasal drug delivery turns out to be a promising routes of delivery and competes with pulmonary drug which is also showing great potential. One of the important points is the almost complete bioavailability and precision of dosage.

2.5.4 Nasal Drug Delivery to the CNS

It is generally believed that drugs pass from the nasal cavity to the CSF via the olfactory epithelium, thus bypassing the blood–brain barrier (BBB). The olfactory nerve is the target when direct absorption into the brain is the goal, because this is the only site in the human body where the CNS is directly expressed on the nasal mucosal surface [4].

2.6 Colorectal Drug Delivery

Although drug administration to the rectum in human beings dates back to 1500 BC, majority of pharmaceutical consumers are reluctant to administer drugs directly by this route. However, the colon is a suitable site for the safe and slow absorption of drugs which are targeted at the large intestine or designed to act systematically. Although the colon has a lower absorption capacity than the small intestine, ingested materials remain in the colon for much longer time. Food passes through the small intestine within a few hours, but it remains in the colon for 2–3 days. Basic requirements of drug delivery to the colorectal area are:

- The drug should be delivered to the colon either in a slow release or targeted form ingested orally or introduced directly by an enema or rectal suppository.
- The drug must overcome the physical barrier of the colonic mucous.
- Drugs must survive metabolic transformation by numerous bacterial species resident in the colon which are mainly anaerobes and possess a wide range of enzymatic activities.

2.6.1 Factors Which Influence Drug Delivery to Colorectal Area

- The rate of absorption of drugs from the colon is influenced by the rate of blood flow to and from the absorptive epithelium.
- Dietary components such as complex carbohydrates trap molecules within polysaccharide chains.
- Lipid-soluble molecules are readily absorbed by passive diffusion.
- The rate of gastric emptying and small bowel transit time.
- Motility patterns of the colon determine the rate of transit through the colon and hence the residence time of a drug and its absorption.
- Drug absorption varies according to whether the drug is targeted to the upper colon, the lower colon, or the rectum.

2.6.2 Advantages of the Rectal Route for Drug Administration

- A relatively large amount of the drug can be administered.
- Oral delivery of drugs that are destroyed by the stomach acid and/or metabolized by pancreatic enzymes.
- This route is safe and convenient particularly for the infants and the elderly.
- This route is useful in the treatment of emergencies in infants such as seizures when the intravenous route is not available.
- The rate of drug absorption from the rectum is not influenced by ingestion of food or rate of gastric emptying.
- The effect of various adjuvants is generally more effective in the rectum than in the upper part of the gastrointestinal tract.

- Drugs absorbed from the lower part of the rectum bypass the liver.
- Degradation of the drugs in the rectal lumen is much less than in the upper gastrointestinal tract.

2.6.3 Disadvantages of the Rectal Route for Drug Administration

- Some hydrophilic drugs such as antibiotics and peptide drugs are not easily absorbed from the rectum, and absorption enhancers are required.
- Drug may cause rectal irritation and sometimes proctitis with ulceration and bleeding.

Drugs targeted for action in the colon can also be administered orally. Colon targeting is recognized to have several therapeutic advantages, such as the oral delivery of drugs that are destroyed by the stomach acid and/or metabolized by pancreatic enzymes. Sustained colonic release of drugs can be useful in the treatment of nocturnal asthma, angina, and arthritis. Local treatment of colonic pathologies, such as ulcerative colitis, colorectal cancer, and Crohn's disease, is more effective with the delivery of drugs to the affected area. Site-specific delivery of an oral anticancer drug to colon for treatment of colorectal cancer increases its concentration at the target site and thus requires a lower dose with reduced incidence of side effects [5]. Likewise, colonic delivery of vermicides and colonic diagnostic agents requires smaller doses.

2.7 Pulmonary Drug Delivery

Although aerosols of various forms for treatment of respiratory disorders have been in use since the middle of twentieth century, the interest in the use of pulmonary route for systemic drug delivery is recent. Interest in this approach has been further stimulated by the demonstration of potential utility of lung as a portal for entry of peptides and the feasibility of gene therapy for cystic fibrosis. It is important to understand the mechanism of macromolecule absorption by the lungs for an effective use of this route.

2.7.1 Mechanisms of Macromolecule Absorption by the Lungs

The lung takes inhaled breaths of air and distributes them deep into the tissue to a very large surface, known as the alveolar epithelium, which is approximately 100 m^2 in adults. This very large surface has approximately a half billion tiny air sacs known as alveoli which are enveloped by an equally large capillary network. The delivery of inhaled air to the alveoli is facilitated by the airways which start with the single trachea and branch several times to reach the grapelike clusters of tiny alveoli. The alveolar volume is 4,000–6,000 ml as compared to airway volume of 400 ml, thus providing a greater area for absorption for the inhaled substances. Large molecule drugs, such as peptides and proteins, do not easily pass through the airway surface because it is lined with a thick, ciliated mucus-covered cell layer making it nearly impermeable. The alveoli, on the other hand, have a thin single cellular layer enabling absorption into the bloodstream. Some barriers to the absorption of substances in the alveoli are:

- Surfactant, a thin layer at the air/water interface and may trap the large molecules.

- A molecule must traverse the surface lining fluid which is a reservoir for the surfactant and contains many components of the plasma as well as mucous.

- The single layer of epithelial cells is the most significant barrier.

- The extracellular space inside the tissues and the basement membrane to which the epithelial cells are attached.

- The vascular endothelium which is the final barrier to systemic absorption is more permeable to macromolecules than the pulmonary epithelium.

Although the mechanism of absorption of macromolecules by the lungs is still poorly understood, the following mechanisms are considered to play a part:

1. Transcytosis (passage through the cells). This may occur and may be receptor mediated, but it is not very significant for macromolecules >40 kDa.

2. Paracellular absorption. This is usually thought to occur through the junctional complex between two cells. The evidence for this route of absorption is not very convincing in case of the lungs. Molecules smaller than 40 kDa may enter via the junctional pores.

3. Drug transporter proteins of the lung epithelium play a role in pulmonary drug delivery, e.g., transporter-dependent absorption of β2-agonists in respiratory epithelial cells [6].

Once past the epithelial barrier, the entry of macromolecules into the blood is easier to predict. Venules and lymph vessels provide the major pathway for absorption. Direct absorption may also occur across the tight junctions of capillary endothelium.

2.7.2 Pharmacokinetic of Inhaled Therapeutics for Systemic Delivery

An accurate estimation of pharmacokinetics of inhaled therapeutics for systemic delivery is a challenging experimental task. Various models for in vivo, in vitro, and ex vivo study of lung absorption and disposition for inhaled therapeutic molecules have been described. In vivo methods in small rodents continue to be the mainstay of assessment as it allows direct acquisition of pharmacokinetic data by reproducible dosing and control of regional distribution in the lungs through the use of different methods of administration. In vitro lung epithelial cell lines provide an opportunity to study the kinetics and mechanisms of transepithelial drug transport in more detail. The ex vivo model of the isolated perfused lung resolves some of the limitations of in vivo and in vitro models. While controlling lung regional distributions, the preparation alongside a novel kinetic modeling analysis enables separate determinations of kinetic descriptors for lung absorption and nonabsorptive clearances, i.e., mucociliary clearance, phagocytosis, and/ or metabolism. There are advantages and disadvantages of each model, and scientists must make appropriate selection of the best model at each stage of the research and development program before proceeding to clinical trials for future inhaled therapeutic entities for systemic delivery.

2.7.3 Advantages of Pulmonary Drug Delivery

- Large surface area available for absorption.
- Close proximity to blood flow.
- Avoidance of first-pass hepatic metabolism.
- Smaller doses are required than by the oral route to achieve equivalent therapeutic effects.

2.7.4 Disadvantages of Pulmonary Drug Delivery

- The lungs have an efficient aerodynamic filter which must be overcome for effective drug deposition to occur.
- The mucous lining the pulmonary airways clears the deposited particles toward the throat.
- Only 10–40 % of the drug leaving an inhalation device is usually deposited in the lungs by using conventional devices.

2.7.5 Techniques of Systemic Drug Delivery via the Lungs

Drugs may be delivered to the lungs for local treatment of pulmonary conditions, but here the emphasis is on the use of the lungs for systemic drug delivery. Simple inhalation devices have been used for inhalation anesthesia, and aerosols containing various drugs have been used in the past. The current interest in delivery of peptides and proteins by this route has led to the use of dry powder formulations for deposition in the deep lung which requires placement within the tracheal bronchial tree rather than simple aerosol inhalation. Various technologies that are in development for systemic delivery of drugs by pulmonary route are:

Dry powders. For many drugs, more active ingredients can be contained in dry powders than in liquid forms. In contrast to aqueous aerosols, where only 1–2 % of the aerosol particle is drug (the rest is water), dry powder aerosol particles can contain up to 50–95 % of pure drug. This means that therapeutic doses of most drugs can be delivered as a dry powder aerosol in one to three puffs. Dry powder aerosols can carry approximately five times more drug in a single breath than metered-dose inhaler (MDI) systems and many more times than currently marketed liquid or nebulizer systems. It is possible that a dry powder system for drugs requiring higher doses, such as insulin or α1-antitrypsin, could decrease dosing time as compared with nebulizers. For example, delivery of insulin by nebulizer requires many more puffs per dose, e.g., up to 50–80 per dose in one study of diabetics. A final reason for focusing on dry powders concerns the microbial growth in the formulation. The risk of microbial growth, which can cause serious lung infections, is greater in liquids than solids.

Inhalers. Various aerosols can deliver liquid drug formulations. The liquid units are inserted into the device which generates the aerosol and delivers it directly to the patient. This avoids any problems associated with converting proteins into powders. This method has applications in delivery of morphine and insulin.

Nanoparticle formulations for pulmonary drug delivery. Nanoparticulate formulations significantly improve drug delivery to the deep lung as well as improve bioavailability. A technique of spray-drying nanoemulsion produces nanoparticles (<100 nm) that can be dispersed homogeneously in the propellant to form an extremely stable pressurized metered-dose inhaler formulation [7]. Methods of preparation and characterization of drug-loaded polymeric nanoparticles prepared from biodegradable charged-modified branched polyesters, aerosolization of the nanosuspensions using a vibrating mesh nebulizer, and evaluation of the pulmonary pharmacokinetics of the nanoscale drug delivery vehicles following aerosol delivery to the airspace of an isolated lung model have been described elsewhere in this book [8].

Controlled-release pulmonary drug delivery. This is suitable for drug agents that are intended to be inhaled, either for local action in the lungs or for systemic absorption. Potential applications for controlled release of drugs delivered through the lungs are:

- It enables reduction of dosing frequency for drugs given several times per day.
- It increases the half-life of drugs which are absorbed very rapidly into the blood circulation and are rapidly cleared from the blood.
- An inhaled formulation may lead to the development of products that might otherwise be abandoned because of unfavorable pharmacokinetics.
- Pulmonary controlled release could decrease development cycles for drug molecules by obviating the need for chemical modification.

2.7.6 Conclusions and Future Prospects of Pulmonary Drug Delivery

The pulmonary route for drug administration is now established for systemic delivery of drugs. A wide range of drugs can be administered by this route but the special attraction is for the delivery of peptides and proteins. Considering the growing number of peptide and protein therapeutic products, several biotechnology companies will get involved in this area. Advances in the production of dry powder formulation will be as important as design of devices for delivery of drugs to the lungs. Effervescent carrier particles can be synthesized with an adequate particle size for deep lung deposition. This opens the door for future research to explore this technology for delivery of a large range of substances to the lungs with possible improved release compared to conventional carrier particles.

Issues of microparticle formation for lung delivery will become more critical in the move from chemically and physically robust small particles to more sensitive and potent large molecules. In spite of these limitations, pulmonary delivery of biopharmaceuticals is an achievable and worthwhile goal. Nanoparticles are being investigated for pulmonary drug delivery, and the current results are promising. Many of the safety issues have been addressed.

Drugs other than biotherapeutics are being developed for inhalation and include treatments either on the market or under development to reduce the symptoms of influenza, to minimize nausea and vomiting following cancer chemotherapy, and to provide vaccinations. Future applications could find inhalable forms of antibiotics to treat directly lung diseases like tuberculosis with large, local doses. Or medications known to cause stomach upsets could be packaged for inhalation, including migraine pain medications, erythromycin, or antidepressants. Inhalable drugs hold the possibility of eliminating common side effects of oral dosages, including low solubility, interactions with food, and low bioavailability. Because inhalables reach the bloodstream faster than pills and some injections, many medical conditions could benefit from fast-acting therapies, including pain, spasms, anaphylaxis, and seizures.

The medicine cabinet of the future may hold various types of inhalable drugs that will replace not only dreaded injections but also drugs with numerous side effects when taken orally. New approaches will lend support to the broad challenge of delivering biotherapeutics and other medications to the lungs.

2.8 Cardiovascular Drug Delivery

Drug delivery to the cardiovascular system is different from delivery to other systems because of the anatomy and physiology of the vascular system; it supplies blood and nutrients to all organs of the body. Drugs can be introduced into the vascular system for systemic effects or targeted to an organ via the regional blood supply. In addition to the usual formulations of drugs such as controlled release, devices are used as well. A considerable amount of cardiovascular therapeutics, particularly for major and serious disorders, involves the use of devices. Some of these may be implanted by surgery, whereas others are inserted via minimally invasive procedures involving catheterization. The use of sophisticated cardiovascular imaging systems is important for the placement of devices. Drug delivery to the cardiovascular system is not simply formulation of drugs into controlled-release preparation, but it includes delivery of innovative therapeutics to the heart [9]. Details of cardiovascular drug delivery are described elsewhere [10].

Methods for local administration of drugs to the cardiovascular system include the following:

- Drug delivery into the myocardium: direct intramyocardial injection, drug-eluting implanted devices
- Drug delivery via coronary venous system
- Injection into coronary arteries via cardiac catheter
- Intrapericardial drug delivery
- Release of drugs into arterial lumen from drug-eluting stents

2.9 Drug Delivery to the Central Nervous System

The delivery of drugs to the brain is a challenge in the treatment of central nervous system (CNS) disorders. The major obstruction to CNS drug delivery is the blood–brain barrier (BBB), which limits the access of drugs to the brain substance. In the past, the treatment of CNS disease was mostly by systemically administered drugs. This trend continues. Most CNS disorder research is directed toward the discovery of drugs and formulations for controlled release; little attention has been paid to the method of delivery of these drugs to the brain. Various methods of delivering drugs to the CNS are shown in Table 2 and are described in detail elsewhere [11–13].

2.9.1 Drug Delivery Across the BBB

Crossing the BBB remains the biggest challenge in drug delivery to the CNS. Nanotechnology-based strategies for crossing the BBB have made significant contribution to this area [14]. One chapter in this book deals with the development of BBB model systems to identify, characterize, and validate novel nanoparticles applicable to brain delivery in vitro [15]. In this work, the authors describe a method to screen nanoparticles with variable size and surface functionalization in order to define the physicochemical characteristics underlying the design of nanoparticles able to efficiently cross the BBB.

Direct introduction of pain medications for action on the spinal cord via the lumbar intrathecal route in human patients is described in another chapter [16]. Crossing of BBB is not required in this approach.

2.10 Concluding Remarks on Routes of Drug Delivery

A comparison of common routes of drug delivery is shown in Table 3. Due to various modifications of techniques, the characteristics can be changed from those depicted in this table. For example, injections can be needle-less and do not have the discomfort leading to better compliance.

Table 2
Various methods of drug delivery to the central nervous system

Systemic administration of therapeutic substances for CNS action

Intravenous injection for targeted action in the CNS

Direct administration of therapeutic substances to the CNS

Introduction into cerebrospinal fluid pathways: intraventricular and subarachnoid pathways such as intrathecal

Introduction into the cerebral arterial circulation

Introduction into the brain substance

Direct positive pressure infusion

Drug delivery by manipulation of the blood–brain barrier

Drug delivery using novel formulations

Conjugates

Gels

Liposomes

Microspheres

Nanoparticles

Chemical delivery systems

Drug delivery devices

Pumps

Catheters

Implants releasing drugs

Use of microorganisms for drug delivery to the brain

Bacteriophages for brain penetration

Bacterial vectors

Cell therapy

CNS implants of live cells secreting therapeutic substances

CNS implants of encapsulated genetically engineered cells producing therapeutic substances

Cells for facilitating crossing of the blood–brain barrier

Gene transfer

Direct injection into the brain substance

Intranasal instillation for introduction into the brain along the olfactory tract

Targeting of CNS by retrograde axonal transport

Vectors: viral and nonviral

© Jain PharmaBiotech

Table 3
Comparison of major routes of drug delivery for systemic absorption

Issue	Oral	Intravenous	Intramuscular/subcutaneous	Transnasal	Transdermal	Pulmonary
Delivery to blood circulation	Indirect through GI tract	Direct	Indirect absorption from tissues	Indirect	Indirect	Indirect
Onset of action	Slow	Rapid	Moderate to rapid	Rapid	Moderate to rapid	Rapid
Bioavailability	Low to high	High	High	Moderate	Low	Moderate to high
Dose control	Moderate	Good	Moderate	Moderate to good	Poor	Moderate to good
Administration	Self	Health professional	Self or health professional	Self	Self	Self
Patient convenience	High	Low	Low	High	Moderate	High
Adverse effects	Gastrointestinal upset	Acute reactions	Acute reactions	Insignificant	Skin irritation	Insignificant
Use for proteins and peptides	No	Yes	Yes	Yes	No	Yes

© Jain PharmaBiotech

3 Drug Formulations

There is constant evolution of the methods of delivery, which involves modifications of conventional methods and discovery of new devices. Some of the modifications of drugs and the methods of administration will be discussed in this section. A classification of technologies that affect the release and availability of drugs is shown in Table 4.

3.1 Sustained Release

Sustained-release (SR) preparations are not new, but several new modifications are being introduced. They are also referred to as "long-acting" or "delayed-release" as compared to "rapid"- or "conventional"-release preparations. The term sometimes overlaps with "controlled release" which implies more sophisticated control of release and not just confined to the time dimension. Controlled release implies consistency, but release of drug in SR preparations may not be consistent. The rationales of developing SR are:

- To extend the duration of action of the drug
- To reduce the frequency of dosing
- To minimize the fluctuations in plasma level
- To improve drug utilization
- To lessen adverse effects

Table 4
Classification of DDS that affect the release and availability of drugs

Systemic versus localized drug delivery
General nontargeted delivery to all tissues
Targeted delivery to a system or organ
Controlled-release delivery systems (systemic delivery)
Release on time scale
Immediate release
Programmed release at a defined time/pulsatile release
Delayed, sustained or prolonged release, long acting
Targeted release (see also drug delivery devices)
Site-specific controlled release following delivery to a target organ
Release in response to requirements or feedback
Receptor-mediated targeted drug delivery
Type of drug delivery device

© Jain PharmaBiotech

Limitations of SR products are:

- Increase of drug cost.

- Variation in the drug-level profile with food intake and from one subject to another.

- The optimal release form is not always defined, and multiplicity of SR forms may confuse the physician as well as the patient.

- SR is achieved by either chemical modification of the drug or modifying the delivery system, e.g., use of a special coating to delay diffusion of the drug from the system. Chemical modification of drugs may alter such properties as distribution, pharmacokinetics, solubility, or antigenicity. One example of this is attachment of polymers to the drugs to lengthen their lifetime by preventing cells and enzymes from attacking the drug.

3.2 Controlled Release

Controlled release implies regulation of the delivery of a drug usually by a device. The control is aimed at delivering the drug at a specific rate for a definite period of time independent of the local environments. The periods of delivery are usually much longer than in the case of SR and vary from days to years. Controlled release may also incorporate methods to promote localization of drug at an active site. Site-specific and targeted delivery systems are the descriptive terms used to denote this type of control.

3.3 Programming the Release at a Defined Time

Approaches used for achieving programmed or pulsatile release may be physical mechanisms such as swelling with bursting or chemical actions such as enzymatic degradation. Capsules have been designed that burst after a predetermined exposure to an aqueous environment. Physical factors that can be controlled are the radius of the sphere, osmotic pressure of the contents, and wall thickness as well as elasticity. Various pulsatile-release methods for oral drug delivery include the Port system (a semipermeable capsule containing an osmotic charge and an insoluble plug) and Chronset system (an osmotically active compartment in a semipermeable cap).

3.4 Prodrugs

A prodrug is a pharmacologically inert form of active drug that must undergo transformation to the parent compound in vivo by either a chemical or an enzymatic reaction to exert its therapeutic effect. The following are required for a prodrug to be useful for site-specific delivery:

- Prodrug must have adequate access to its pharmacological receptors.

- The enzyme or chemical responsible for activating the drug should be active only at the target site.

Table 5
Novel preparations for improving bioavailability of drugs

Oral drug delivery
Fast-dissolving tablets
Technologies to increase gastrointestinal retention time
Technologies to improve drug-release mechanisms of oral preparations
Adjuvants to enhance absorption
Methods of increasing bioavailability of drugs
Penetration enhancement
Improved dissolution rate
Inhibition of degradation prior to reaching site of action
Production of therapeutic substances inside the body
Gene therapy
Cell therapy

© Jain PharmaBiotech

- The enzyme should be in adequate supply to produce the required level of the drug to manifest its pharmacological effects.
- The active drug produce at the target site should be retained there and not diffuse into the systemic circulation.

Example of a prodrug is L-dopa, the precursor of dopamine, which is distributed systemically following oral administration. Its conversion to dopamine in the corpus striatum of the brain produces the desired therapeutic effects.

3.5 Novel Carriers and Formulations for Drug Delivery

Various novel methods of delivery have evolved since the simple administration of pills and capsules as well as injections. These involve formulations as shown in Table 5 and carriers as shown in Table 6. Biodegradable implants are shown in Table 7.

3.6 Antibody-Targeted Systems

Drug delivery systems can make use of macromolecular attachment for delivery using immunoglobulins as the macromolecule. The obvious advantage of this system is that it can be targeted to the site of antibody specificity. The advantages are that less amount of the drug is required and the side effects are reduced considerably.

Drugs are linked, covalently or non-covalently to the antibody, or placed in vesicles such as liposomes or microspheres, and the antibody is used to target the liposome. Covalent attachments are generally not very efficient and diminish the antigen-binding capacity. If conjugation is done through an intermediate carrier

Table 6
Novel carriers for drug delivery

Polymeric carriers for drug delivery
Collagen
Particulate drug delivery systems: microspheres
Nanobiotechnology-based methods including nanoparticles such as liposomes
Glass-like sugar matrices
Resealed red blood cells
Antibody-targeted systems: radiolabeled antibodies

© Jain PharmaBiotech

Table 7
Biodegradable implants for controlled sustained drug delivery

Injectable implants
Gels
Microspheres
Surgical implants
Sheets/films
Foams
Scaffolds

© Jain PharmaBiotech

molecule, it is possible to increase the drug/antibody ratio. Such intermediates include dextran or poly-L-glutamic acid.

Examples of drugs that have been conjugated to antibodies or their fragments are anticancer drugs. Numerous antibody–liposome combinations have been investigated for delivering drugs and genes. The term immunoliposomes is used for liposomes loaded with drug cargo that have been surface-conjugated to antibodies. The main advantage of antibody-targeted system is that the adverse effects of anticancer drugs can be reduced by the use of monoclonal antibodies that recognize only tumor antigens.

Radiolabeled antibodies are used in research as well as clinically for imaging and therapeutic purposes. They provide an opportunity for combining diagnosis with drug delivery. Because of their ability to selectively target tumor antigens, radiolabeled monoclonal antibodies are used as direct therapeutic agents for cancer radioimmunotherapy. One chapter describes methods for labeling of antibodies with radioiodine and radiometals [17].

3.7 Gene Therapy as a Drug Delivery Method

Gene therapy can be broadly defined as the transfer of defined genetic material to specific target cells of a patient for the ultimate purpose of preventing or altering a particular disease state [18]. Gene therapy is an efficient method of delivery of therapeutics, and other delivery methods are used to deliver genes.

Certain bacteria with natural tumor specificity have been used as gene vectors for specifically delivering genes or gene products to the tumor environment following intravenous administration by a process termed "bactofection." One chapter describes procedures for studying this phenomenon in vitro and in vivo for both invasive and noninvasive bacteria that can be exploited as tumor-specific therapeutic delivery vehicles for the transfer of plasmid DNA to mammalian cells in rodent models [19].

A listing of various technologies involved in gene therapy gives an idea of the broad scope as shown in Table 8. Controlled and targeted gene therapy is promising.

3.7.1 Delivery of siRNAs

Successful silencing of genes requires efficient delivery of siRNAs. Various methods of delivery of siRNA have been used including targeted delivery by lipid-based technologies. siRNA–lipid nanoparticles (siRNA–LNP) have potential therapeutic applications, and one chapter describes a novel microfluidic-based manufacturing process for the rapid manufacture of siRNA–LNP, together with protocols for characterizing the size, polydispersity, RNA encapsulation efficiency, RNA concentration, and total lipid concentration of the resultant nanoparticles [20].

3.8 Cells as Vehicles for Drug Delivery

Cells are used as drug delivery vehicles. Drugs encapsulated in red blood cell (RBC), cell-based gene therapy, and implantation of encapsulated cells secreting therapeutic proteins are examples of these.

3.8.1 Cell-Mediated Gene Therapy

Transplantation of cells with specific functions is a recognized procedure for the treatment of human diseases because cells can function as biological pumps for release of therapeutic substances. Cell-mediated gene therapy involves the genetic manipulation of cells followed by their in vitro amplification and subsequent injection of the modified cells into target tissues. Several types of cells have been used for cell-mediated gene therapy.

The ability of stem cells to home to sites of acute injury could serve as a means of local drug delivery through the infusion of genetically engineering stem cells that secrete gene products of interest. Mesenchymal stem cells (MSCs) have been shown to migrate toward and engraft into the tumor sites, which provide a potential for their use as carriers for cancer gene therapy using a nonviral transfection method as described in one of the chapters [21].

Table 8
Methods of gene therapy

Gene transfer

Chemical: calcium phosphate transfection

Physical

Electroporation

Gene gun

Transduction with recombinant virus vectors

Adeno-associated virus

Adenovirus

Herpes simplex virus

Lentivirus

Moloney murine leukemia virus

Retroviral vectors

Vaccinia virus

Other viruses

Nonviral vectors for gene therapy

Liposomes

Ligand–polylysine–DNA complexes

Dendrimers and other polycationic polymers

Synthetic peptide complexes

Artificial viral vectors

Artificial chromosomes

Use of microorganisms as oncolytic agents

Bacteria for gene delivery

Viral oncolysis

Cell/gene therapy

Administration of cells modified ex vivo

Implantation of genetically engineered cells to produce therapeutic substances

Gene/DNA administration

Direct injection of naked DNA or genes: systemic or at target site

Receptor-mediated endocytosis

Use of refined methods of drug delivery, e.g., microspheres

(continued)

Table 8
(continued)

Gene regulation

Regulation of expression of delivered genes in target cells by locus control region technology

Light-activated gene therapy

Molecular switch to control expression of genes in vivo

Promoter element-triggered gene therapy

Repair of defective genes

Involves correction of the gene in situ, e.g., chimeraplasty

Gene repair mediated by single-stranded oligonucleotides

Gene replacement

Excision or replacement of the defective gene by a normal gene

Spliceosome-mediated RNA trans-splicing

Inhibition of gene expression

Antisense oligodeoxynucleotides

Antisense RNA

Ribozymes

RNA interference: delivery of small interfering RNAs (siRNAs)

© Jain PharmaBiotech

3.8.2   RBCs as Drug Delivery Vehicles

RBCs represent naturally designed carriers for intravascular drug delivery, characterized by unique longevity in the bloodstream, biocompatibility, and safe physiological mechanisms for metabolism [22]. Several protocols of infusion of RBC-encapsulated drugs are being explored in patients. Delivery of drugs, particularly those targeting phagocytic cells and those that must act within the vascular lumen, may benefit from carriage by RBCs. Two strategies for RBC drug delivery are (1) encapsulation into isolated RBCs ex vivo followed by infusion in compatible recipients and (2) coupling of drugs to the surface of RBCs, e.g., those regulating immune response. RBC drug delivery by injection of therapeutics conjugated with fragments of antibodies provides safe anchoring of cargoes to circulating RBC, without the need for ex vivo modification and infusion of RBC.

When RBCs are placed in a hypotonic medium, they swell with rupture of the membrane and formation of pores. This allows encapsulation of 25 % of the drug or enzyme in solution.

Table 9
Methods of delivery of cells for therapeutic purposes

Injection
Subcutaneous
Intramuscular
Intravenous
Intrathecal
Implantation into various organs by surgical procedures, e.g., brain, spinal cord, myocardium
Oral intake of encapsulated cells
Pharmacologically active microcarriers
Use of special devices for delivery of cells
Cell delivery systems to promote growth of cells for regenerative medicine

© Jain PharmaBiotech

The membrane is resealed by restoring the tonicity of the solution. Potential uses of loaded RBCs as drug delivery systems are:

- They are biodegradable and non-immunogenic.
- They can be modified to change their resident circulation time; depending on their surface, cells with little surface damage can circulate for a longer time.
- Entrapped drug is shielded from immunological detection and external enzymatic degradation.
- The system is relatively independent of the physicochemical properties of the drug.

The drawbacks of using RBCs are that the damaged RBCs are sequestered in the spleen and the storage life is limited to about 2 weeks.

3.8.3 Drug Delivery Systems for Cell Therapy

Like gene therapy vectors, cells may deliver therapeutics, but there is also a need for drug delivery systems for cell and gene therapies. Various methods of delivery of cells for therapeutic purposes are listed in Table 9.

An important objective in cell therapy for regenerative medicine is delivery of materials to promote growth of cells. One chapter in this book describes the development of a self-assembling peptide hydrogel and its potential use as a cell and growth factor delivery vehicle to the infarcted heart in a rodent model of myocardial infarction [23].

3.9 Ideal Properties of Material for Drug Delivery

The properties of an ideal macromolecular drug delivery or biomedical vector are:

- Structural control over size and shape of drug or imaging-agent cargo space
- Biocompatible, nontoxic polymer/pendant functionality
- Precise, nanoscale container and/or scaffolding properties with high drug or imaging-agent capacity features
- Well-defined scaffolding and/or surface modifiable functionality for cell-specific targeting moieties
- Lack of immunogenicity
- Appropriate cellular adhesion, endocytosis, and intracellular trafficking to allow therapeutic delivery or imaging in the cytoplasm or nucleus
- Acceptable bio-elimination or biodegradation
- Controlled or triggerable drug release
- Molecular level isolation and protection of the drug against inactivation during transit to target cells
- Minimal nonspecific cellular and blood-protein binding properties
- Ease of consistent, reproducible, clinical-grade synthesis

3.10 Innovations for Improving Oral Drug Delivery

Oral route remains the most common method of drug delivery. Efforts are constantly made to made this route more efficient and faster.

3.10.1 Fast-Dissolving Tablets

Fast-disintegration technology is used for manufacturing these tablets. The advantages of fast-dissolving tablets are:

- Convenient to take without the use of water
- Easier to take by patients who cannot swallow
- Rapid onset of action due to faster absorption
- Less gastric upset because the drug is dissolved before it reaches the stomach
- Improved patient compliance

3.10.2 Softgel Formulations

Capsules and other protective coatings have been used to protect the drugs in their passage through the upper gastrointestinal tract for delayed absorption. The coatings also serve to reduce stomach irritation. The softgel delivers drugs in solution and yet offers advantages of solid dosage form. Softgel capsules are particularly suited for hydrophobic drugs which have poor bioavailability because these drugs do not dissolve readily in water and gastrointestinal juices. If hydrophobic drugs are compounded in solid dosage forms, the dissolution rate may be slow, absorption is variable, and the bioavailability is incomplete. Bioavailability is improved in the presence of fatty acids, e.g., mono- or diglycerides. Fatty acids can solubilize hydrophobic drugs such as hydrochlorothiazide, isotretinoin, and griseofulvin in the gut and facilitate rapid absorption. Hydrophobic drugs are dissolved in hydrophilic solvent and encapsulated. When softgels are crushed or chewed, the drug is released immediately in the gastric juice and is absorbed from the gastrointestinal tract into the bloodstream. This results in rapid onset of desired therapeutic effects. Advantages of softgels over tablets are:

- The development time for softgel is shorter due to lower bioavailability concerns, and such solutions can be marketed at a fraction of cost.

- Softgel formulations, e.g., that of ibuprofen, have a shorter time to peak plasma concentration and greater peak plasma concentration compared to a marketed tablet formulation. Cyclosporin can give therapeutic blood levels which are not achievable from tablet form. Similarly oral hypoglycemic glipizide in softgel is also known to have better bioavailability results compared with tablet form.

- Softgel delivery systems can also incorporate phospholipids or polymers or natural gums to entrap the drug active in the gelatin layer with an outer coating to give desired delayed-/controlled-release effects.

Advantages of softgel capsule over other hard-shell capsules are:

- Sealed tightly in automatic manner.
- Easy to swallow.
- Allow product identification, using colors and several shapes.
- Better stability than other oral delivery systems.
- Good availability and rapid absorption.
- Offer protection against contamination, light, and oxidation.
- Unpleasant flavors are avoided due to content encapsulation.

3.10.3 Improving Drug-Release Mechanisms of Oral Preparations

Drug-release rates of orally administered products tend to decrease from the matrix system as a function of time based on the nature and method of preparation. Various approaches to address the problems associated with drug-release mechanisms and release rates use geometric configurations including the cylindrical rod method and the cylindrical donut methods. The three-dimensional printing (3DP) provides the following advantages:

- Zero-order drug delivery
- Patterned diffusion gradient by microstructure diffusion barrier technique
- Cyclic drug release

Table 10
Classification of drug delivery devices

Surgically implanted devices for prolonged sustained drug release
Drug reservoirs
Surgically implanted devices for controlled/intermittent drug delivery
Pumps and conduits
Implants for controlled release of drugs (nonbiodegradable)
Implantable biosensor drug delivery system
Microfluidic device for drug delivery
Controlled-release microchip
Implants that could benefit from local drug release
Vascular stents: coronary, carotid, and peripheral vascular
Ocular implants
Dental implants
Orthopedic implants

© Jain PharmaBiotech

3DP method utilizes ink-jet printing technology to create a solid object by printing a binder into selected areas of sequentially deposited layers of powder. The active agent can be embedded into the device as either dispersion along the polymeric matrix or as discrete units in the matrix structure. The drug-release mechanism can be tailored for a variety of requirements such as controlled release by a proper selection of polymer material and binder material.

3.11 Drug Delivery Devices

One of the most obvious ways to provide sustained-release medication is to place the drug in a delivery device and implant the system into body tissue. A classification of drug delivery devices is shown in Table 10.

The concept of drug delivery devices is old, but the new technologies are being applied. Surgical techniques and special injection devices are sometimes required for implantation. The materials used for these implants must be biocompatible, i.e., the polymers used should not cause any irritation at the site of implantation or promote an abscess formation. Subcutaneous implantation is currently one of the utilized routes to investigate the potential of sustained delivery systems. Favorable absorption sites are available, and the device can be removed at any time.

A variety of other drugs have been implanted subcutaneously including thyroid hormones, cardiovascular agents, insulin, and nerve growth factor. Some implantable devices extend beyond simple sources of drug diffusion. Some devices can be triggered by changes in osmotic pressure to release insulin, and pellets can be activated by magnetism to release their encapsulated drug load. Such external control of an embedded device would eliminate many of the disadvantages of most implantable drug delivery systems.

3.11.1   Pumps and Conduits for Drug Delivery

Mechanical pumps are usually miniature devices such as implantable infusion pumps and percutaneous infusion catheters which deliver drugs into appropriate vessels or other sites in the body. Several pumps, implantable catheters, and infusion devices are available commercially. Examples of applications of these devices are:

- Intrathecal morphine infusion for pain control
- Intraventricular drug administration for disorders of the brain
- Hepatic arterial chemotherapy
- Intravenous infusion of heparin in thrombotic disorders
- Intravenous infusion of insulin in diabetes

The advantages of these devices are:

- The rate of drug diffusion can be controlled.
- Relatively large amounts of drugs can be delivered.
- The drug administration can be changed or stopped when required.

3.11.2   Controlled-Release Microchip for Drug Delivery

Microchip technology has been applied to achieve pulsatile release of liquid solutions. A solid-state silicon microchip, which incorporates micrometer-scale pumps and flow channels, can provide controlled release of single or multiple chemical substances on demand. The release mechanism is based on the electrochemical dissolution of thin anode membranes covering microreservoirs filled with chemicals in various forms. Varying amounts of chemical substances in solid, liquid, or gel form can be released in either a pulsatile or a continuous manner or a combination of both. The entire device can be mounted on the tip of a small probe or implanted in the body.

Microchip Technology Inc. uses its technology for controlled-release iontophoretic transdermal device for drug delivery, which can use a smaller battery size and a lighter weight battery and reduce overall size and weight. Only small footprint microcontrollers are needed for wearable patch-based iontophoresis devices. The conventional controlled drug release from polymeric materials is in response to specific stimuli such as electric and magnetic fields, ultrasound, light, enzymes, etc.

Continuous glucose monitoring has been associated with improved glycemic control in adults with type 1 diabetes. Microchip Technology Inc. is developing a long-term implanted continuous glucose monitor designed to wirelessly deliver continuous, convenient, reliable, and accurate glucose measurements to guide the delivery of insulin. The company has also successfully conducted a clinical trial of this device for delivery of PTH1-34 for the treatment of osteoporosis [24].

Drug delivery based on lab-on-a-chip technology is evolving with incorporation of advances in micro- and nanotechnologies. Main lab-on-a-chip drug delivery systems in development are (1) a concentration gradient generator integrated with a cell culture platform at the cellular level, (2) synthesis of smart particles as drug carriers at the tissue level, and (3) microneedles and implantable devices with fluid-handling components at the organism level [25].

In the future, proper selection of a biocompatible material may enable the development of an autonomous controlled-release implant which has been dubbed as "pharmacy-on-a-chip" or a highly controlled tablet (smart tablet) for drug delivery. The researchers hope to engineer the chips so that they can change the drug-release schedule or medication type in response to commands beamed through the skin. Each microchip contains an array of discrete reservoirs from which dose delivery can be controlled by telemetry.

3.11.3   Use of Physical Agents for Controlled Release of Therapeutic Substances

There are several examples of use of physical agents such as heat and ultrasound for controlling release of therapeutic agents. One example described in a chapter in this book is a delivery system using noninvasive, nonviral mediated method of gene therapy using ultrasound for site-specific gene delivery [26]. This novel platform technology uses gas-filled, acoustic microspheres for both diagnostic imaging and therapy and may provide a key component for future success in the pursuit of single gene replacement therapy.

3.12   Particulate Drug Delivery Systems

The concept of using particles to deliver drugs to selected sites of the body originated from their use as radiodiagnostic agents in medicine in the investigation of the reticuloendothelial system (liver, spleen, bone marrow, and lymph nodes). Particles ranging from 20 µm up to 300 µm have been proposed for drug targeting. Because of the small size of the particles, they can be injected directly into the systemic circulation or a certain compartment of the body. Particulate drug delivery systems may contain an intimate mixture of the drug, and the core material or the drug may be dispersed as an emulsion in the carrier material, or the drug may be encapsulated by the carrier material. Factors which influence the release of drugs from particulate carriers are:

- The drug: Its physicochemical properties, position in the particle, and drug–carrier interaction
- Particles: Type, size, and density of the particle
- Environments: Temperature, light, presence of enzymes, ionic strength, and hydrogen ion concentration

Various particulate drug carrier systems can be grouped into the following classes:

- Microspheres are particles larger than 1 μm but small enough not to sediment when suspended in water (usually 1–100 μm).
- Nanoparticles are colloidal particles ranging in size between 10 and 1,000 nm.
- Glass-like sugar matrices.
- Liposomes.
- Cellular particles such as resealed erythrocytes, leukocytes, and platelets.

3.12.1 Microspheres

Microspheres prepared from cross-linked proteins have been used as biodegradable drug carriers. The rate of release of small drug molecules from protein microspheres is relatively rapid, although various strategies such as complexing the drug with macromolecules can be adopted to overcome this problem. Polysaccharides (e.g., starch) and a wide range of synthetic polymers have been used to manufacture microspheres. Microcapsules differ from microspheres in having a barrier membrane surrounding a solid or liquid core which is an advantage in the case of peptides and proteins. Special applications of microspheres and microcapsules are:

- Poly-D,,L-lactide-*co*-glycolide-agarose microspheres can encapsulate protein and stabilize them for drug delivery.
- Multicomponent, environmentally responsive, hydrogel microspheres, coated with a lipid bilayer can be used to mimic the natural secretory granules for drug delivery.
- Microencapsulation of therapeutic agents to provide local controlled drug release in the central nervous system across the blood–brain barrier.
- Microspheres can be used for chemoembolization of tumors in which the vasculature is blocked while anticancer agent is released from the trapped microparticles.
- Microcapsules, produced at ideal size for inhalation (1–5 μm), can be used in formulating drugs for pulmonary delivery, both for local delivery and for systemic absorption.
- Microspheres can be used as nasal drug delivery systems for systemic absorption of peptides and proteins.
- Poly-D,L-lactide-*co*-glycolide microspheres can be used as a controlled-release antigen delivery system—parenteral or oral.
- Delivery of antisense oligonucleotides.
- Nano-encapsulation of DNA in bioadhesive particles can be used for gene therapy by oral administration.

3.12.2 Glass-Like Sugar Matrices

These are microparticles made of glass-like sugar matrix. The solution of sugar and insulin is sprayed as a mist into a stream of hot, dry air which quickly dries the mist to a powder, a process known as spray drying. The transformation from liquid to a glassy powder is rapid and prevents denaturation of the insulin. Sugar microspheres can also be used for preserving drugs and vaccines which normally require refrigeration for travel to remote parts of the world. Sugar molecules protect the drug molecules by "propping up" the active structure, preventing it from denaturing when the water molecules are removed.

3.12.3 Resealed Red Blood Cells

Red blood cells (RBCs) have been studied the most of all the cellular drug carriers. When RBCs are placed in a hypotonic medium, they swell with rupture of the membrane and formation of pores. This allows encapsulation of 25 % of the drug or enzyme in solution. The membrane is resealed by restoring the tonicity of the solution. Potential uses of loaded RBCs as drug delivery systems are:

- They are biodegradable and non-immunogenic.
- They can be modified to change their resident circulation time; depending on their surface, cells with little surface damage can circulate for a longer time.
- Entrapped drug is shielded from immunological detection and external enzymatic degradation.
- The system is relatively independent of the physicochemical properties of the drug.

The drawbacks of using RBCs are that the damaged RBCs are sequestered in the spleen and the storage life is limited to about 2 weeks.

3.13 Nanotechnology-Based Drug Delivery

Nanotechnology is the creation and utilization of materials, devices, and systems through the control of matter on the nanometer-length scale, i.e., at the level of atoms, molecules, and supramolecular structures. It is the popular term for the construction and utilization of functional structures with at least one characteristic dimension measured in nanometer—a nanometer is one billionth of a meter (10^{-9} m). Nanotechnologies are described in detail in a special report on this topic [27].

Trend toward miniaturization of carrier particles had already started prior to the introduction of nanotechnology in drug delivery. The suitability of nanoparticles for use in drug delivery depends on a variety of characteristics, including size and porosity. Nanoparticles can be used to deliver drugs to patients through various routes of delivery. Nanoparticles are important for delivering drugs intravenously so that they can pass safely through the body's smallest blood vessels, for increasing the surface area of a drug so that it will dissolve more rapidly, and for delivering drugs via inhalation. Porosity is important for entrapping gases in nanoparticles, for controlling the release rate of the drug, and for targeting drugs to specific regions.

It is difficult to create sustained-release formulations for many hydrophobic drugs because they release too slowly from the nanoparticles used to deliver the drug, diminishing the efficacy of the delivery system. Modifying water uptake into the nanoparticles can speed the release, while retaining the desired sustained-release profile of these drugs. Water uptake into nanoparticles can be modified by adjusting the porosity of the nanoparticles during manufacturing and by choosing from a wide variety of materials to include in the shell.

Nanobiotechnology provides the following solutions to the problems of drug delivery:

- Improving solubilization of the drug.

- Using noninvasive routes of administration eliminates the need for administration of drugs by injection.

- Development of novel nanoparticle formulations with improved stabilities and shelf lives.

- Development of nanoparticle formulations for improved absorption of insoluble compounds and macromolecules enable improved bioavailability and release rates, potentially reducing the amount of dose required and increasing safety through reduced side effects.

- Manufacture of nanoparticle formulations with controlled particle sizes, morphology, and surface properties would be more effective and less expensive than other technologies.

- Nanoparticle formulations that can provide sustained-release profiles up to 24 h can improve patient compliance with drug regimens.

- Direct coupling of drugs to targeting ligand restricts the coupling capacity to a few drug molecules, but coupling of drug carrier nanosystems to ligands allows import of thousands of drug molecules by means of one receptor-targeted ligand. Nanosystems offer opportunities to achieve drug targeting with newly discovered disease-specific targets.

3.13.1   Nanomaterials and Nanobiotechnologies Used for Drug Delivery

Various nanomaterials and nanobiotechnologies used for drug delivery are shown in Table 11.

3.13.2   Liposomes

Liposomes are stable microscopic vesicles formed by phospholipids and similar amphipathic lipids. Liposome properties vary substantially with lipid composition, size, surface charge, and the method

Table 11
Nanomaterials and nanobiotechnologies used for drug delivery

Structure	Size	Role in drug delivery
Bacteriophage NK97 (a virus that attacks bacteria)	~28 nm	Emptied of its own genetic material, NK97, which is covered by 72 interlocking protein rings, can act as a nanocontainer to carry drugs and chemicals to targeted locations
Canine parvovirus (CPV)-like particles	~26 nm	Targeted drug delivery: CPV binds to transferrin receptors, which are overexpressed in some tumors
Carbon magnetic nanoparticles	40–50 nm	For drug delivery and targeted cell destruction
Ceramic nanoparticles	~35 nm	Accumulate exclusively in the tumor tissue and allow the drug to act as sensitizer for PDT without being released
Cerasomes	60–200 nm	Cerasome is filled with C6 ceramide for use as an anticancer agent
Dendrimers	1–20 nm	Holding therapeutic substances such as DNA in their cavities
Gold nanoparticles	2–4 nm	Enable externally controlled drug release
HTCC nanoparticles	110–180 nm	Encapsulation efficiency is up to 90 %. In vitro release studies show a burst effect followed by a slow and continuous release
Micelle/nanopill	25–200 nm	Made from two polymer molecules—one water repellant and the other hydrophobic—that self-assemble into a sphere called a micelle that can deliver drugs to specific structures within the cell
Low-density lipoproteins	20–25 nm	Drugs solubilized in the lipid core or attached to the surface
Nanocochleates		Nanocochleates facilitate delivery of biologicals such as genes
Nanocrystals	<1,000 nm	Nanocrystal technology (Elan) can rescue a significant number of poorly soluble chemical compounds by increasing solubility
Nanodiamonds	550–800 nm	Biocompatibility and unique surface properties for drug delivery
Nanoemulsions	20–25 nm	Drugs in oil and/or liquid phases to improve absorption
Nanoliposomes	25–50 nm	Incorporate fullerenes to deliver drugs that are not water soluble and tend to have large molecules
Nanoparticle composites	~40 nm	Attached to guiding molecules such as MAbs for targeted drug delivery
Nanopore membrane		An implanted titanium device using silicon nanopore membrane can release encapsulated protein and peptide drugs
Nanospheres	50–500 nm	Hollow ceramic nanospheres created by ultrasound

(continued)

Table 11
(continued)

Structure	Size	Role in drug delivery
Nanosponges	10 nm	A long, linear molecule scrunched into a sphere ~10 nm in ⌀ with a large number of surface sites for drug molecule attachment
Nanostructured organogels	~50 nm	Mixture of olive oil and liquid solvents and adding a simple enzyme to chemically activate a sugar. Used to encapsulate drugs
Nanotubes	Single wall 1–2 nm Multiwall 20–60 nm	Resemble tiny drinking straws and that might offer advantages over spherical nanoparticles for some applications
Nanovalve	~500 nm	Externally controlled release of drug into a cell
Nanovesicles	25–100 nm	Bilayer spheres containing the drugs in lipids
Niosomes	300–500 nm	Nonionic surfactant-based liposomes that are more stable and have better penetrating capability than liposomes
Polymer nanocapsules	50–200 nm	Enclosing drugs
PEG-coated PLA nanoparticles	Variable size	PEG coating improves the stability of PLA nanoparticles in the gastrointestinal fluids and helps the transport of encapsulated protein across the intestinal and nasal mucus membranes
Quantum dots	2–10 nm	Combine imaging with therapeutics
Superparamagnetic iron oxide nanoparticles	10–100 nm	As drug carriers for intravenous injection to evade RES as well as to penetrate the very small capillaries within the body tissues and thus offer the most effective distribution in certain tissues

Abbreviations: *PEG* poly(ethylene glycol), *PLA* poly(lactic acid), *HTCC* N-(2-hydroxyl) propyl-3-trimethyl ammonium chitosan chloride, *RES* reticuloendothelial system

© Jain PharmaBiotech

of preparation. They are therefore divided into three classes based on their size and number of bilayers:

1. Small unilamellar vesicles are surrounded by a single lipid layer and are 25–50 nm in diameter.

2. Large unilamellar vesicles are a heterogeneous group of vesicles similar to and are surrounded by a single lipid layer.

3. Multilamellar vesicles consist of several lipid layers separated from each other by a layer of aqueous solution.

Lipid bilayers of liposomes are similar in structure to those found in living cell membranes and can carry lipophilic substances such as drugs within these layers in the same way as cell membranes. The pharmaceutical properties of the liposomes depend on the composition of the lipid bilayer and its permeability and fluidity. Cholesterol, an important constituent of many cell membranes, is frequently included in liposome formulations because it reduces the permeability and increases the stability of the phospholipid bilayers.

Until recently, the use of liposomes as therapeutic vectors was hampered by their toxicity and lack of knowledge about their biochemical behavior. The simplest use of liposomes is as vehicles for drugs and antibodies targeted for the targeted delivery of anti-cancer agents. The use of liposomes may be limited because of problems related to stability, the inability to deliver to the right site, and the inability to release the drug when it gets to the right site. However, liposome surfaces can be readily modified by attaching polyethylene glycol (PEG) units to the bilayer (producing what is known as stealth liposomes) to enhance their circulation time in the bloodstream. Furthermore, liposomes can be conjugated to antibodies or ligands to enhance target-specific drug therapy.

Liposome-based strategies have been used for innovative drug delivery technologies that involve crossing various membranes to reach the target site of action. An innovative MITO-Porter system described in another chapter of this book has been developed for overcoming the mitochondrial membrane via membrane fusion for use in delivering a wide variety of carrier-encapsulated molecules, regardless of size or physicochemical properties, to mitochondria [28]. To deliver cargos to target organelles using this strategy, the liposomes must fuse with the organelle membrane.

3.13.3 Polymer Nanoparticles

Biodegradable polymer nanoparticles are poly(ethylene glycol) (PEG)-coated poly(lactic acid) (PLA) nanoparticles, chitosan (CS)-coated poly(lactic acid–glycolic acid) (PLGA) nanoparticles, and chitosan (CS) nanoparticles. These nanoparticles can carry and deliver proteins in an active form and transport them across the nasal and intestinal mucosa. Additionally, PEG coating improves the stability of PLA nanoparticles in the gastrointestinal fluids and helps the transport of the encapsulated protein, tetanus toxoid, across the intestinal and nasal mucous membranes. Furthermore, intranasal administration of these nanoparticles provided high and long-lasting immune responses.

Thermosensitive polymers where temperature is used as a triggering signal can be used for controlled and targeted drug delivery. Synthesis and characterization of a thermoresponsive polymer, its application for drug delivery in bovine serum albumin drug model, and evaluation of cytotoxicity of the polymers are described in detail in another chapter of this book [29].

3.13.4 Chitosan Nanoparticles

Chitosan (CS) is a biodegradable natural polymer derived from chitin. Authors of one chapter have evaluated the efficacy of chitosan nanoparticles of different sizes as drug delivery systems for anticancer drugs and found that chitosan-100 kDa was the more effective drug carrier than the chitosan-15 and chitosan-200 KD [30].

N-(2-hydroxyl) propyl-3-trimethyl ammonium chitosan chloride is water-soluble derivative of CS, synthesized by the reaction between glycidyl-trimethyl ammonium chloride and CS. Coating of PLGA nanoparticles with the mucoadhesive CS improves the stability of the particles in the presence of lysozyme and enhanced the nasal transport of the encapsulated tetanus toxoid. Nanoparticles made solely of CS are stable upon incubation with lysozyme. Moreover, these particles are very efficient in improving the nasal absorption of insulin as well as the local and systemic immune responses to tetanus toxoid, following intranasal administration. Chitosan–siRNA nanoparticle formulation may be introduced directly in the brain by intranasal delivery as described elsewhere in this book [31].

3.13.5 Polymeric Micelles

Micelles that are biocompatible nanoparticles varying in size from 50 to 200 nm in which poorly soluble drugs can be encapsulated represent a possible solution to the delivery problems associated with such compounds. Such micelles could be exploited to target the drugs to particular sites in the body, potentially alleviating toxicity problems. pH-sensitive drug delivery systems can be engineered to release their contents or change their physicochemical properties in response to variations in the acidity of the surroundings.

3.13.6 Nanosponge for Drug Delivery

Nanosponges are hyper-cross-linked cyclodextrin polymers nanostructured to form 3D networks and are obtained by complexing cyclodextrin with a cross-linker such as carbonyldiimidazole. Another method for making a nanosponge uses extensive internal cross-linking to scrunch a long, linear molecule into a sphere about 10 nm in diameter. Instead of trying to encapsulate drugs in nanoscale containers, this approach creates a nanoparticle with a large number of surface sites where drug molecules can be attached. A molecular transporter attached to the nanosponge can carry it and its cargo across biological barriers into specific intracellular compartments. Nanosponges have been used to increase the solubility and stability of poorly soluble drugs. β-Cyclodextrin nanosponges loaded with anticancer agent tamoxifen have been used for oral drug delivery [32]. In experimental studies, tamoxifen nanosponge complex with particle size of 400–600 nm was shown to be more cytotoxic than plain tamoxifen after 24 and 48 h of incubation.

3.13.7 Future Prospects of Nanotechnology-Based Drug Delivery

A desirable situation in drug delivery is to have smart drug delivery systems that can integrate with the human body. This is an area where nanotechnology will play an extremely important role. Even time-release tablets, which have a relatively simple coating that dissolves in specific locations, involve the use of nanoparticles. Pharmaceutical companies are already involved in using nanotechnology to create intelligent drug-release devices. For example, control of the interface between the drug/particle and the human body can be programmed so that when the drug reaches its target, it can then become active. The use of nanotechnology for drug-release devices requires autonomous device operation. For example, in contrast to converting a biochemical signal into a mechanical signal and being able to control and communicate with the device, autonomous device operation would require biochemical recognition to generate forces to stimulate various valves and channels in the drug delivery systems, so that it does not require any external control.

3.14 Receptor-Mediated Targeted Drug Delivery

Receptor-mediated endocytosis is a process whereby extracellular macromolecules and particles gain entry to the intracellular environments. Cell surface receptors are complex transmembrane proteins which mediate highly specific interactions between cells and their extracellular environment. The cells use receptor-mediated endocytosis for nutrition, defense, transport, and processing. Cellular uptake of drugs bound to a targeting carrier or to a targetable polymeric carrier is mostly restricted to receptor-mediated endocytosis. Because receptors are differentially expressed in various cell types and tissues, using receptors as markers may be an advantageous strategy for drug delivery. Receptor-mediated uptake can achieve the specific transport of the drug to the receptor-bearing target cells. Many receptors such as receptors for transferrin, low-density lipoprotein, and asialoglycoprotein have been used to deliver drugs to specific types of cells or tissues.

Many recent advances in targeted drug delivery have focused on the regulation of the endogenous membrane trafficking machinery in order to facilitate uptake of drugs via receptor-mediated endocytosis into target tissues. Vesicle motor proteins (kinesin, cytoplasmic dynein, and myosin) play an important role in membrane trafficking events and are referred to as molecular motors. It is important to understand the events involved in the movement of surface-bound and extracellular components by endocytosis into the cell. Knowledge of sorting events within the endocytic pathways that govern the intracellular destination and the ultimate fate of the drug is also important. If internalization of the drug is followed by recycling or degradation, no accumulation can occur within the cell. Strategies for regulation of such events can enhance drug delivery. One example is the delivery of genes via receptor-mediated endocytosis in hepatocytes. Factors that influence the efficiency of the receptor-mediated uptake of targeted drug conjugate are:

- Affinity of the targeting moieties
- The affinity and nature of target antigen
- Density of the target antigen

- The type of cell targeted
- The rate of endocytosis
- The route of internalization of the receptor–ligand complex
- The ability of the drug to escape from the vesicular compartment into the cytosol
- The affinity of the carrier to the drug

Receptor-mediated drug delivery is particularly applicable to cytotoxic therapy for cancer and gene therapy.

3.15 Targeted Delivery Systems

For targeted and controlled delivery, a number of carrier systems and homing devices are under development: glass-like matrices, monoclonal antibodies, resealed erythrocytes, microspheres, and liposomes. There are more sophisticated systems based on molecular mechanisms, nanotechnology, and gene delivery. These will be discussed in the following pages.

3.15.1 Polymeric Carriers for Drug Delivery

The limitation of currently available drug therapies, particularly for the treatment of diseases localized to specific organs, has led to efforts to develop alternative methods of drug administration to increase their specificity. One approach for this purpose is the use of degradable polymeric carriers for drugs which are delivered to and deposited at the site of the disease for extended periods with minimal systemic distribution of the drug. The polymeric carrier is degraded and eliminated from the body shortly after the drug has been released. The polymers are divided into three groups:

1. Nondegradable polymers. These are stable in biological systems. These are mostly used as components of implantable devices for drug delivery.
2. Drug-conjugated polymers. In these, the drug is attached to a water-soluble polymer carrier by a cleavable bond. These polymers are less accessible to healthy tissues as compared with the diseased tissues. These conjugates can be used for drug targeting via systemic administration or by implanting them directly at the desired site of action where the drug is released by cleavage of the drug–polymer bond. Examples of such polymers are dextrans, polyacrylamides, and albumin.
3. Biodegradable polymers. These degrade under biological conditions to nontoxic products that are eliminated from the body.

Macromolecular complexes of various polymers can be divided into the following categories according to the nature of molecular interactions:

- Complexes formed by interaction of oppositely charged polyelectrolytes
- Charge transfer complexes

- Hydrogen-bonding complexes
- Stereocomplexes

Polyelectrolyte complexes can be used as implants for medical use, as microcapsules, or for binding of pharmaceutical products including proteins. In recent years, a new class of organometallic polymers, polyphosphazenes, have become available. Synthetic flexibility of polyphosphazenes makes them a suitable material for controlled-release technologies. Desirable characteristics of a polymeric system used for drug delivery are:

- Minimal tissue reaction after implantation
- High polymeric purity and reproducibility
- A reliable drug-release profile
- In vivo degradation at a well-defined rate in case of biodegradable implants

Polymeric delivery systems for implanting at specific sites are either a reservoir type where the drug is encapsulated into a polymeric envelope that serves as a diffusional rate-controlled membrane or a matrix type where the drug is evenly dispersed in a polymer matrix. Most of the biodegradable systems are of the matrix type where the drug is released by a combination of diffusion, erosion, and dissolution. Disadvantages of the implants are that once they are in place, the dose cannot be adjusted and the discontinuation of therapy requires a surgical procedure to remove the implant. For chronic long-term release, repeated implantations are required.

The development of injectable biodegradable drug delivery systems has provided new opportunities for controlled drug delivery as they have advantages over traditional ones such as ease of application and prolonged localized drug delivery. Both natural and synthetic polymers have been used for this purpose. Following injection in fluid state, they solidify at the desired site. These systems have been explored widely for the delivery of various therapeutic agents ranging from antineoplastic agents to proteins and peptides such as insulin. Polymers are also being used as nanoparticles for drug delivery as described in Subheading 3.13.

3.15.2 Evaluation of Polymers In Vivo

Biodegradable polymers have attracted much attention as implantable drug delivery systems. Uncertainty in extrapolating in vitro results to in vivo systems due to the difficulties of appropriate characterization in vivo, however, is a significant issue in the development of these systems. To circumvent this limitation, non-electron paramagnetic resonance (EPR) and magnetic resonance imaging (MRI) have been applied to characterize drug release and polymer degradation in vitro and in vivo. MRI makes it possible to monitor water content, tablet shape, and response of the biological system such as edema and encapsulation. The results of the MRI experiments give the first direct proof in vivo of postulated mechanisms of polymer erosion. Using nitroxide radicals as model drug-releasing compounds, information on the mechanism of drug release and microviscosity inside the implant can be obtained by means of EPR spectroscopy. The use of both noninvasive methods to monitor processes in vivo leads to new insights in understanding the mechanisms of drug release and polymer degradation.

3.15.3   Collagen

Collagen, being a major protein of connective tissues in animals, is widely distributed in the skin, bones, teeth, tendons, eyes, and most other tissues in the body and accounts for about one-third of the total protein content in mammals. It also plays an important role in the formation of tissues and organs and is involved in various cells in terms of their functional expression. Collagen as a biomaterial has been used for repair and reconstruction of tissues and as an agent for wound dressing.

Several studies have already been conducted on the role of collagen as a carrier in drug delivery. In vivo absorption of collagen is controlled by the use of a cross-linking agent such as glutaraldehyde or by induction of cross-linking through ultraviolet or gamma ray irradiation in order to enhance the sustained-release effects. Release rate of drugs can be controlled by (1) collagen gel concentration during preparation of the drug delivery system, (2) the form of drug delivery system, and (3) the degree of cross-linking of the collagen.

3.15.4   Avidin–Biotin Systems for Targeted Drug Delivery

Avidin, originally isolated from chicken eggs, and its bacterial analogue, streptavidin, from *Streptomyces avidinii*, have extremely high affinity for biotin. This unique feature is the basis of avidin–biotin technology. The current status of avidin–biotin systems and their use for pretargeted drug delivery and vector targeting have been reviewed elsewhere [33]. Some of the useful applications are:

- Targeting antibodies and therapeutic molecules are administered separately leading to a reduction of drug dose in normal tissues compared with conventional (radio)immunotherapies.
- Introducing avidin gene into specific tissues by local gene transfer, which subsequently can sequester and concentrate considerable amounts of therapeutic ligands.
- Enables transductional targeting of gene therapy vectors.

3.15.5   Targeted Delivery of Biologicals with Controlled Release

Targeted delivery to the organ or site or lesion in the body is an important part of therapeutics. This is particularly important in case of biologicals, and delivery is often combined with controlled release of the drug or therapeutic molecules by the use of intrinsic or extrinsic mechanisms. One chapter in this book describes the protocol for synthesis of a smart drug delivery system based on gold nanoparticles, which has been shown to release the bioactive material in response to an intracellular stimulus, i.e., glutathione concentration gradient [34].

3.16 Methods of Administration of Proteins and Peptides

Various possible routes for administration of proteins and peptides are:

- Parenteral
- Transdermal
- Inhalation
- Transnasal
- Oral
- Rectal
- Implants
- Cell and gene therapies
- Use of special formulations

Injection still remains the most common method for administration of proteins and peptides. Efforts are being made to use needle-free or painless injections and also to improve the controlled delivery by parenteral route.

3.16.1 Delivery of Peptides by Subcutaneous Injection

Subcutaneous still remains predictable and controllable route of delivery for peptides and macromolecules. However, there is need for greater convenience and lower cost for prolonged and repeated delivery. An example of refinement of subcutaneous delivery is MEDIPAD (Elan Pharmaceutical Technologies) which is a combination of "patch" concept and a sophisticated miniaturized pump operated by gas generation. It was described in the report on transdermal drug delivery.

3.16.2 Depot Formulations and Implants

These are usually administered by injection and must ensure protein/peptide stability. One of the formulations used is poly(lactide-co-glycolide) sustained release. Example of an approved product in the market is leuprolide. Implants involve invasive administration and also must ensure protein/peptide stability. Implantable titanium systems provide drug release driven by osmotic pumps. This technology has been extended to other proteins such as growth hormone. Nutropin Depot (Genentech/Alkermes) was the first long-acting form of growth hormone that encapsulates the drug in biodegradable microspheres that release the hormone slowly after injection. It reduces the frequency of injection in children with growth hormone deficiency from once daily to once a month.

3.16.3 Poly(ethylene Glycol) Technology

Poly(ethylene glycol) or PEG, a water-soluble polymer, is a well-recognized treatment for constipation. When covalently linked to proteins, PEG alters their properties in ways that extend their potential uses. Chemical modification of proteins and other bioactive molecules with PEG—a process referred to as PEGylation—can be used to tailor molecular properties to particular applications, eliminating disadvantageous properties or conferring new molecular functions. This approach can be used to improve delivery of proteins and peptides.

---

4 New Concepts in Pharmacology That Influence Design of DDS

Pharmacology, particularly pharmacokinetics and pharmacodynamics, has traditionally influenced drug delivery formulations. Some of the newer developments in pharmacology and therapeutics that influence the development of DDSs are:

1. Pharmacogenetics
2. Pharmacogenomics
3. Pharmacoproteomics
4. Pharmacometabolomics
5. Chronopharmacology

The first four items are linked together and form the basis of personalized medicine, which will be discussed later in this chapter.

4.1 Pharmacogenetics

Pharmacogenetics, a term recognized in pharmacology in the pregenomic era, is the study of influence of genetic factors on the action of drugs as opposed to genetic causes of disease. Now it is the study of the linkage between the individual's genotype and the individual's ability to metabolize a foreign compound. The pharmacological effect of a drug depends on pharmacodynamics (interaction with the target or the site of action) and pharmacokinetics (absorption, distribution, and metabolism). It also covers the influence of various factors on these processes. Drug metabolism is one of the major determinants of drug clearance and the factor that is most often responsible for interindividual differences in pharmacokinetics.

The differences in response to medications are often greater among members of a population than they are within the same person or between monozygotic twins at different times. The existence of large population differences with small intrapatient variability is consistent with inheritance as a determinant of drug response. It is estimated that genetics can account for 20–95 % of variability in drug disposition and effects. Genetic polymorphisms in drug-metabolizing enzymes, transporters, receptors, and other drug targets have been linked. From this initial definition, the scope has broadened so that it overlaps with pharmacogenomics. Genes influence pharmacodynamics and pharmacokinetics. Pharmacogenetics has a threefold role in the pharmaceutical industry, which is relevant to the development of personalized medicines:

1. For study of the drug metabolism and pharmacological effects
2. For predicting genetically determined adverse reactions
3. Drug discovery and development and as an aid to planning clinical trials

4.2 Pharmacogeno-mics

Pharmacogenomics, a distinct discipline within genomics, carries on the tradition by applying the large-scale systemic approaches of genomics to understand the basic mechanisms and apply them to drug discovery and development. Pharmacogenomics now seeks to examine the way drugs act on the cells as revealed by the gene expression patterns and thus bridges the fields of medicinal chemistry and genomics. Some of the drug response markers are examples of interplay between pharmacogenomics and pharmaco-genetics; both are playing an important role in the development of personalized medicines [35]. The two terms—pharmacogenetics and pharmacogenomics—are sometimes used synonymously, but one must recognize the differences between the two.

Various technologies enable the analysis of these complex multifactorial situations to obtain individual genotypic and gene expression information. These same tools are applicable to study the diversity of drug effects in different populations. Pharmacogenomics promises to enable the development of safer and more effective drugs by helping to design clinical trials such that nonresponders would be eliminated from the patient population and take the guesswork out of prescribing medications. It will also ensure that the right drug is given to the right person from the start. In clinical practice, doctors could test patients for specific single nucleotide polymorphisms (SNPs) known to be associated with nontherapeutic drug effects before prescribing in order to determine which drug regimen best fits their genetic makeup. Pharmacogenomic studies are rapidly elucidating the inherited nature of these differences in drug disposition and effects, thereby enhancing drug discovery and providing a stronger scientific basis for optimizing drug therapy on the basis of each patient's genetic constitution.

Pharmacogenomics provides a new way of looking at the old problems, i.e., how to identify and target the essential component of disease pathway(s). These changes will increase the importance of drug delivery systems which need to be adapted to our changing concept of the disease. Drug delivery problems have to be considered parallel to all stages of drug development from discovery to clinical use.

4.3 Pharmacopro-teomics

The term "proteomics" indicates PROTEins expressed by a genOME and is the systematic analysis of protein profiles of tissues. There is an increasing interest in proteomic technologies now because deoxyribonucleic acid (DNA) sequence information provides only a static snapshot of the various ways in which the cell might use its proteins, whereas the life of the cell is a dynamic process. Role of proteomics in drug development can be termed "pharmacoproteomics." Proteomics-based characterization of multifactorial diseases may help to match a particular target-based therapy to a particular marker in a subgroup of patients. The industrial sector is taking a lead in developing this area. Individualized therapy may be based on differential protein expression rather than a genetic polymorphism.

4.4   Chronopharma-cology

The term "chronopharmacology" is applied to variations in the effect of drugs according to the time of their administration during the day. Mammalian biological functions are organized according to circadian rhythms (lasting about 24 h). They are coordinated by a biological clock situated in the suprachiasmatic nuclei (SCN) of the hypothalamus. These rhythms persist under constant environmental conditions, demonstrating their endogenous nature. Some rhythms can be altered by disease. The rhythms of disease and pharmacology can be taken into account to modulate treatment over the 24-h period.

The knowledge of such rhythms appears particularly relevant for the understanding and/or treatment of hypertension and ischemic coronary artery disease. In rats and in man, the circadian rhythm of systolic or diastolic blood pressure can be dissociated from the rest–activity cycle, suggesting that it is controlled by an oscillator which can function independently of the SCN, which could justify modification of treatment according to the anomalies of the blood pressure rhythm. The morning peak of myocardial infarction in man is due to the convergence of several risk factors, each of which has a 24-h cycle: blood coagulability, BP, oxygen requirements, and myocardial susceptibility to ischemia. The existence of these rhythms and the chronopharmacology of cardiovascular drugs such as nitrate derivatives constitute clinical prerequisites for the chronopharmacotherapy of heart disease.

It is known that the sensitivity of tumor cells to chemotherapeutic agents can depend on circadian phase. There are possible differences in rhythmicity of cells within tissues. If cells within a tumor are not identically phased, this may allow some cells to escape from the drug's effect. Perhaps synchronizing the cells prior to drug treatment would improve tumor eradication. Wild-type and circadian mutant mice demonstrate striking differences in their response to the anticancer drug cyclophosphamide. While the sensitivity of wild-type mice varies greatly, depending on the time of drug administration, Clock mutant mice are highly sensitive to treatment at all times tested. These findings will provide a rationale not only for adjusting the timing of chemotherapeutic treatment to be less toxic but also for providing a basis for a search for pharmacological modulators of drug toxicity acting through circadian system regulators. This result may significantly increase the therapeutic index and reduce morbidity associated with anticancer treatment.

Chronopharmacological drug formulations can provide the optimal serum levels of the drug at the appropriate time of the day or night. For example, if the time of action desired is early morning, drug release is optimized for that time, whereas with conventional methods of drug administration, the peak will be reached in the earlier part of the night and with controlled release, the patient will have a constant high level throughout the night. Effective chronopharmacotherapeutics will not only improve the efficacy of treatment but will open up new markets. This approach to treatment requires suitable drug delivery systems.

4.5 Impact of Current Trends in Pharmaceutical Product Development on DDS

Considerable advances have taken place in pharmaceutical industry during the past two decades. Contemporary trends in pharmaceutical product development which are relevant to DDS are listed in Table 12. Drug delivery technologies have become an important part of the biopharmaceutical industry. Drug delivery systems, pharmaceutical industry, and biotechnology interact with each other as shown in Fig. 1.

4.6 Impact of New Biotechnologies on Design of DDS

New biotechnologies have a great impact on the design of DDS during the past decade. The most significant of these technologies is nanobiotechnology.

Table 12
Current trends in pharmaceutical product development

Use of recombinant DNA technology
Expansion of use of protein and peptide drugs in current therapeutics
Introduction of antisense, RNA interference and gene therapy
Advances in cell therapy: introduction of stem cells
Miniaturization of drug delivery: microparticles and nanoparticles
Increasing use of bioinformatics and computer drug design
A trend towards development of target organ-oriented dosage forms
Increasing emphasis on controlled-release drug delivery
Use of routes of administration other than injections
Increasing alliances between pharmaceutical companies and DDS companies

© Jain PharmaBiotech

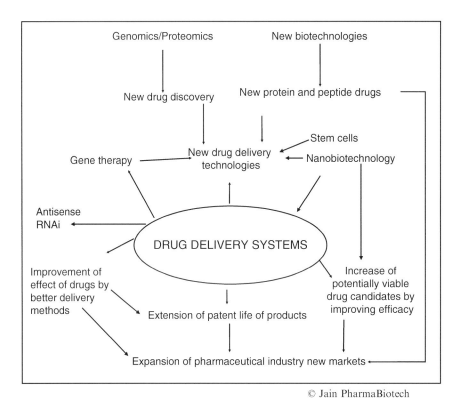

© Jain PharmaBiotech

Fig. 1 Interrelationship of DDS, pharmaceutical industry, and biotechnology. © Jain PharmaBiotech

5   Aims of DDS Development

Drug delivery technologies are aimed at improving efficacy and safety of medicines as well as commercial pharmaceutical development. The important points are:

- Improvement of drug safety and efficacy
- Improved compliance
- Chronopharmacological benefits
- Reduction of cost of drug development
- Life extension of the products
- Reduction of risk of failure in new product development

5.1   Improvement of Safety and Efficacy

Improvement of the safety and efficacy of existing medications is a common objective for all those involved in healthcare. There is no doubt that improved delivery of medications with longer duration of action leads to increased efficacy. Lesser quantities of the active ingredients are required, and targeted application can spare the rest of the body from side effects.

5.2 Improved Compliance

Compliance is a big problem in medical care. Most patients do not like to take medications or fail to take them as instructed. Drugs with sustained release can remedy some of these problems. Once daily dosage with sustained action is likely to improve the compliance rate to 80 % compared to 40 % for three or four times a day. Some of the novel delivery methods such as transdermal or buccal are preferred by most patients to oral intake or injections. Lack of compliance is responsible for a significant number of hospital admission in the USA (as high as 10 % in some estimates). Improvement of compliance can lead to significant reductions in healthcare costs.

5.3 Life Extension of the Products

Patents of several proprietary drugs are about to expire. The introduction of an improved dosage delivery form prior to the expiry of the patent allows the manufacturer to maintain the product with some advantages over generic copies. Several aspects of this are:

- Alternative dosage form may offer advantages over the old product such as improved compliance, increased safety, and enhanced efficacy.

- Development of a previously unknown delivery formulation of an old product would enable a new patent to be obtained even though the active ingredient is the same.

- Sustained-release versions of an older drug are easy to copy in the generic form, but high-tech drug delivery forms are not easy to copy.

- Drugs with expiring patents can be converted into proprietary over-the-counter products to maintain brand franchise.

5.4 Economic Factors

Economics is the most important driver for the development of drug delivery technology. Benefits of new formulations are perceptible at various levels of drug development and patient care as follows:

Continued revenues after expiry of patent. A new drug delivery method can continue to generate revenues for the manufacturer years after expiry of the patent for the original active ingredient. This may sometimes exceed the earnings from the original product.

Market extension. New formulations based on novel drug delivery systems can open up new indications and new markets for the old product. Calcium antagonists were originally launched for angina but achieved more success in the management of hypertension following development of immediate-release formulations.

Drug rescue. Several drugs are discontinued at various stages of development process because of lack of suitable delivery technology. Some promising products get into clinical trials only to be dismissed because of unacceptable toxicity or lack of efficacy. This means considerable financial loss for the companies. An appropriate drug delivery system may rescue some of these products by overcoming these difficulties and also increase the number of potential drugs for clinical trials from the preclinical pipeline of a company.

Reduction of cost of drug development. Compared to the high cost and long development time of a pharmaceutical product, a drug delivery system takes a much shorter time to develop—usually 2–4 years—and costs much less. Development for a new formulation of a generic preparation has more potential of profit in relation to investment as compared with developing a new chemical entity.

Reduction of financial risk. With large investment in R&D of biotechnology products, there is a considerable element of risk involved, and a number of biotechnology companies have failed. In contrast, development of drug delivery products requires only a fraction of capital investment and less time to approval. The use of already approved drugs takes away the element of risk involved in approval of the main ingredient of the DDS.

Reduction of healthcare costs. Lack of compliance is a considerable burden on the healthcare systems because of increased hospitalizations. Improvement of compliance by appropriate drug delivery systems reduces the cost of healthcare. An appropriate delivery system also reduces the amount of drug used and thus reduces the costs.

Competitive advantage. In today's pharmaceutical marketplace, with several products for the same pharmaceutical category, a suitable drug delivery system may help in providing an advantage in competition. A product with a better and more appropriate drug delivery system may move ahead of its competitors if the economic advantages based on improved efficacy, safety, and compliance can be demonstrated.

6   Impact of Current Trends in Healthcare on DDS

Medicine is constantly evolving from the impact of new technologies. In the past, medicine has been more of an art than a science, but the effect of new discoveries in life sciences is having its effect on the practice of medicine. Revolutionary discoveries in molecular biology did not have an immediate impact on medicine, and there is a lag period before changes are noticeable in the practice of medicine. Many of these changes come from better understanding of the disease, whereas others come from improvements in pharmaceuticals and their delivery. One of the most important trends in healthcare is the concept of personalized medicine.

6.1   Personalized Medicine

Personalized medicine simply means the prescription of specific treatments and therapeutics best suited for an individual [36]. It is also referred to as individualized or individual-based therapy. Personalized medicine is based on the idea of using a patient's genotype as a factor in deciding on treatment options, but other factors are also taken into consideration. Genomic/proteomic technologies have facilitated the development of personalized medicines, but other technologies are also contributing to this effort. This process of personalization starts at the development stage of a medicine and is based on pharmacogenomics and pharmacogenetics. Selection of a DDS most appropriate for a patient would be a part of personalized medicine.

Because all major diseases have a genetic component, knowledge of genetic basis helps in distinguishing between clinically similar diseases. Classifying diseases based on genetic differences in affected individuals rather than by clinical symptoms alone makes diagnosis and treatment more effective. Identifying human genetic variations will eventually allow clinicians to subclassify diseases and adapt therapies to the individual patients.

Several diseases can now be described in molecular terms. Some defects can give rise to several disorders, and diseases will be reclassified on molecular basis rather than according to symptoms and gross pathology. The implication of this is that the same drug can be used to treat a number of diseases with the same molecular basis. Another way of reclassification of human diseases will be the subdivision of patient populations within the same disease group according to genetic markers and response to medications.

Along with other technologies, refinements in drug delivery will play an important role in the development of personalized medicine. One well-known example is glucose sensors regulating the release of insulin in diabetic patients. Gene therapy, as a sophisticated drug delivery method, can be regulated according to the needs of individual patients.

7 Characteristics of an Ideal DDS

Characteristics of an ideal drug delivery system are:

- It should increase the bioavailability of the drug.
- It should provide for controlled drug delivery.
- It should transport the drug intact to the site of action while avoiding the non-diseased host tissues.
- The product should be stable, and delivery should be maintained under various physiological variables.
- A high degree of drug dispersion.
- The same method should be applicable to a wide range of drugs.
- It should be easy to administer to the patients.
- It should be safe and reliable.
- It should be cost effective.

8   Drug Delivery Systems for Differential Release of Multiple Drugs

Combination therapy with multiple therapeutic agents is wide applicability in medical and surgical treatment, being used widely, especially in the treatment of cancer. Drug delivery systems that can differentially release two or more drugs are useful for this purpose. New techniques can be used to engineer established drug delivery systems and synthesize new materials as well as to design carriers with new structures for multi-agent delivery systems. A publication has reviewed multi-agent delivery systems from nanoscale to bulk scale, such as liposomes, micelles, polymer conjugates, nano-/microparticles, and hydrogels, developed over the last 10 years [37]. The chemical structure of drug delivery systems is the key to controlling the release of therapeutic agents in combination therapy, and the differential release of multiple drugs could be realized by the successful design of a proper delivery system. Besides biological evaluation in vitro and in vivo, it is important to translate resulting delivery systems into clinical applications.

9   Integration of Diagnostics, Therapeutics, and DDS

Combination of diagnostics with therapeutics is an important component of personalized medicine. A DDS can be integrated into this combination to control the delivery of therapeutics in response to variations in the patient's condition as monitored by diagnostics. Nanobiotechnology has helped in this integration. One example is the use of quantum dots in cancer, where diagnosis, therapeutics, and drug delivery can be combined using QDs as the common denominator [38].

The pharmaceutical industry is taking an active part in the integration of diagnostics and therapeutics. During drug development, there is an opportunity to guide the selection, dosage, route of administration, and multidrug combinations to increase the efficacy and reduce toxicity of pharmaceutical products.

10   Current Achievements, Challenges, and Future of Drug Delivery

Considerable advances have occurred in DDS within the past decade. Extended-release, controlled-release, and once-a-day medications are available for several commonly used drugs. Global vaccine programs are close to becoming a reality with the use of oral, transmucosal, transcutaneous, and needle-less vaccination. Considerable advances have been made in gene therapy and delivery of protein therapeutics. Many improvements in cancer treatment can be attributed to novel drug delivery technologies.

New drug delivery systems will develop during the next decade by interdisciplinary collaboration of material scientists, engineers, biologists, and pharmaceutical scientists. Progress in microelectronics and nanotechnology is revolutionizing drug delivery. However, DDS industry is still facing challenges, and some of these are as follows:

- Drug delivery technologies require constant redesigning to keep up with the new methods of drug design and manufacture, particularly of biotechnology products.

- As the costs of drugs are rising, drug delivery aims to reduce the costs by improving the bioavailability of drugs so that lesser quantities need to be taken in by the patient.

- More fundamental research needs to be done to characterize the physiological barriers to therapy such as the blood–brain barrier.

- New materials that are being discovered, such as nanoparticles, need to have safety studies and regulatory approval.

Future of DDS can be predicted to some extent for the next 10 years. To go beyond that, e.g., 20 years from now, would be very speculative. Some of the drugs currently used would disappear from the market, and no one knows for sure what drugs would be discovered in the future. Some of the diseases may be partially eliminated, and new variants may appear, particularly in infections. With this scenario, it would be difficult to say what methods will evolve to deliver the drugs that we do not know as yet. Some of the drug treatments may be replaced by devices that do not involve the use of drugs.

References

1. Jain KK (2014) Transdermal drug delivery: technologies, markets and companies. Jain PharmaBiotech Publications, Basel, pp 1–280

2. Donnelly RF, Garland MJ, Alkilani AZ (2014) Microneedle-iontophoresis combinations for enhanced transdermal drug delivery, Chapter 7. In: Jain KK (ed) Drug delivery system, Methods Mol Biol. Springer, New York

3. Singh RM, Kumar A, Pathak K (2013) Mucoadhesive in situ nasal gelling drug delivery systems for modulated drug delivery. Expert Opin Drug Deliv 10:115–130

4. Gizurarson S (2012) Anatomical and histological factors affecting intranasal drug and vaccine delivery. Curr Drug Deliv 9:566–582

5. Krishnaiah YS, Khan MA (2012) Strategies of targeting oral drug delivery systems to the colon and their potential use for the treatment of colorectal cancer. Pharm Dev Technol 17: 521–540

6. Salomon JJ, Ehrhardt C (2012) Organic cation transporters in the blood-air barrier: expression and implications for pulmonary drug delivery. Ther Deliv 3(6):735–747

7. Li HY, Zhang F (2012) Preparation of nanoparticles by spray-drying and their use for efficient pulmonary drug delivery. Methods Mol Biol 906:295–301

8. Beck-Broichsitter M, Dalla-Bona AC, Kissel T, Seeger W, Schmehl T (2014) Polymer nanoparticle-based controlled pulmonary drug delivery, Chapter 8. In: Jain KK (ed) Drug delivery system, Methods Mol Biol. Springer, New York

9. Jain KK (2011) Applications of biotechnology in cardiovascular therapeutics. Springer Science, New York

10. Jain KK (2014) Cardiovascular drug delivery: technologies, markets and companies. Jain PharmaBiotech Publications, Basel, pp 1–268

11. Jain KK (ed) (2010) Drug delivery to the central nervous system. Humana, New York

12. Jain KK (2013) Applications of biotechnology in neurology. Springer, New York

13. Jain KK (2014) Drug delivery in CNS disorders: technologies, markets and companies. Jain PharmaBiotech Publications, Basel, pp 1–360

14. Jain KK (2012) Nanobiotechnology-based strategies for crossing the blood-brain barrier. Nanomedicine 7:1225–1233

15. Guarnieri D, Muscetti O, Netti PA (2014) A method for evaluating nanoparticle transport through the blood brain barrier in vitro, Chapter 12. In: Jain KK (ed) Drug delivery system, Methods Mol Biol. Springer, New York

16. De Andres J et al (2014) Intrathecal delivery of analgesics, Chapter 16. In: Jain KK (ed) Drug delivery system, Methods Mol Biol. Springer, New York

17. Gupta S, Batra S, Jain M (2014) Antibody labeling with radioiodine and radiometals, Chapter 9. In: Jain KK (ed) Drug delivery system, Methods Mol Biol. Springer, New York

18. Jain KK (2014) Gene therapy: technologies, markets and companies. Jain PharmaBiotech Publications, Basel, pp 1–665

19. Cummins J, Cronin M, van Pijkeren JP, Gahan C, Tangney M (2014) Bacterial systems for gene delivery to systemic tumours, Chapter 13. In: Jain KK (ed) Drug delivery system, Methods Mol Biol. Springer, New York

20. Cullis PR, Ramsay E (2014) Microfluidic-based manufacture of siRNA-lipid nanoparticles for therapeutic applications, Chapter 6. In: Jain KK (ed) Drug delivery system, Methods Mol Biol. Springer, New York

21. Gao JQ, Hu UL (2014) Stem cells as carriers for cancer gene therapy, Chapter 5. In: Jain KK (ed) Drug delivery system, Methods Mol Biol. Springer, New York

22. Muzykantov VR (2010) Drug delivery by red blood cells: vascular carriers designed by mother nature. Expert Opin Drug Deliv 7:403–427

23. Boopathy AV, Davis ME (2014) Self assembling peptide-based delivery of therapeutics for myocardial infarction, Chapter 10. In: Jain KK (ed) Drug delivery system, Methods Mol Biol. Springer, New York

24. Farra R, Sheppard NF Jr, McCabe L et al (2012) First-in-human testing of a wirelessly controlled drug delivery microchip. Sci Transl Med 4(122):122ra21

25. Nguyen NT, Shaegh SA, Kashaninejad N, Phan DT (2013) Design, fabrication and characterization of drug delivery systems based on lab-on-a-chip technology. Adv Drug Deliv Rev 65(11–12):1403–1419, pii: S0169-409X(13)00127-0

26. Castle J (2014) Ultrasound-directed site-specific gene delivery, Chapter 3. In: Jain KK (ed) Drug delivery system, Methods Mol Biol. Springer, New York

27. Jain KK (2014) Nanobiotechnology: applications, markets and companies. Jain PharmaBiotech Publications, Basel, pp 1–774

28. Yamada Y, Harashima H (2014) Methodology for screening mitochondrial fusogenic envelopes for use in mitochondrial drug delivery, Chapter 2. In: Jain KK (ed) Drug delivery system, Methods Mol Biol. Springer, New York

29. Mishra S, De A, Mozumdar S (2014) Synthesis of a thermoresponsive polymer for drug delivery, Chapter 4. In: Jain KK (ed) Drug delivery system, Methods Mol Biol. Springer, New York

30. Tajmir-Riahi HA, Nafisi S, Sanyakamdhorn S et al (2014) Applications of chitosan nanoparticles in drug delivery, Chapter 11. In: Jain KK (ed) Drug delivery system, Methods Mol Biol. Springer, New York

31. Malhotra M, Tomaro-Duchesneau C, Saha S, Prakash S (2014) Intranasal delivery of chitosan-siRNA nanoparticle formulation to the brain, Chapter 15. In: Jain KK (ed) Drug delivery system, Methods Mol Biol. Springer, New York

32. Torne S, Darandale S, Vavia P et al (2013) Cyclodextrin-based nanosponges: effective nanocarrier for Tamoxifen delivery. Pharm Dev Technol 18:619–625

33. Lesch HP, Kaikkonen MU, Pikkarainen JT, Ylä-Herttuala S (2010) Avidin-biotin technology in targeted therapy. Expert Opin Drug Deliv 7:551–564

34. De A, Mishra S, Garg S, Mozumdar S (2014) Synthesis of a smart nanovehicle for targeting liver, Chapter 14. In: Jain KK (ed) Drug delivery system, Methods Mol Biol. Springer, New York

35. Jain KK (2014) Personalized medicine: scientific and commercial aspects. Jain PharmaBiotech Publications, Basel, pp 1–860

36. Jain KK (2009) A textbook of personalized medicine. Springer Science, New York

37. Zhang H, Wang G, Yang H (2011) Drug delivery systems for differential release in combination therapy. Expert Opin Drug Deliv 8:171–190

38. Jain KK (2014) Drug delivery in cancer: technologies, markets and companies. Jain PharmaBiotech Publications, Basel, pp 1–668

Chapter 2

A Method for Screening Mitochondrial Fusogenic Envelopes for Use in Mitochondrial Drug Delivery

Yuma Yamada and Hideyoshi Harashima

Abstract

Various types of mitochondrial dysfunctions have been implicated in a variety of human diseases. This suggests that mitochondria would be promising therapeutic drug targets and mitochondrial therapy would be expected to be useful for the treatment of various diseases. We have already reported the development of a MITO-Porter, a liposome-based nano-carrier that delivers its cargo to mitochondria via a membrane-fusion mechanism. In our strategy for delivering cargos to mitochondria using a MITO-Porter, the carriers must fuse with the organelle membrane. Here we report on methodology for screening various types of lipid envelopes that have the potential for fusing with a mitochondrial membrane. The method involves monitoring the cancellation of fluorescence resonance energy transfer (FRET) and evaluating membrane fusion between the carriers and mitochondria in living cells by FRET analysis using a spectral imaging fluorescent microscopy system.

Key words Mitochondria, Mitochondrial drug delivery, Mitochondrial macromolecule delivery, MITO-Porter, Membrane fusion, Mitochondrial gene therapy, Mitochondrial medicine, Octaarginine

1 Introduction

Various types of mitochondrial dysfunction have recently been implicated in a variety of human diseases. Therefore, this organelle is a promising therapeutic drug target, and mitochondrial therapy would be expected to be useful and productive for the treatment of various diseases. The targeted delivery of an engineered gene or gene product to the mitochondrion is an essential first step towards the therapeutic restoration of a missing cellular function. The conjugation of a mitochondrial-targeting signal (MTS) peptide to exogenous proteins and small linear DNAs was reported to aid their delivery to mitochondria [1–4], but this strategy is not viable for delivering pDNA. This is because large molecules such as pDNA and folded proteins do not readily pass through the mitochondrial membrane, thus making it nearly impossible for them to be easily delivered to mitochondria [5, 6].

Kewal K. Jain (ed.), *Drug Delivery System*, Methods in Molecular Biology, vol. 1141,
DOI 10.1007/978-1-4939-0363-4_2, © Springer Science+Business Media New York 2014

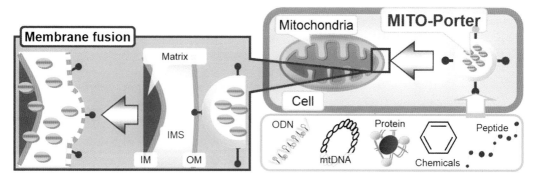

Fig. 1 MITO-Porter is surface-modified with a high density of R8, which can be internalized by cells via macropinocytosis. When the MITO-Porter reaches the cytosol, it binds to mitochondria via electrostatic interactions with R8. Encapsulated compounds are delivered to the intramitochondrial compartment via fusogenic lipids that fuse to the mitochondrial membrane. *OM* outer membrane, *IMS* intermembrane space, *IM* inner membrane [22]

To overcome these problems, we recently proposed an original and innovative strategy for overcoming the mitochondrial membrane via membrane fusion, namely, the development of a MITO-Porter system (Fig. 1) [7–9]. This MITO-Porter is potentially promising for use in delivering a wide variety of carrier-encapsulated molecules, regardless of size or physicochemical properties, to mitochondria. The first barrier to intracellular targeting is the plasma membrane. In a previous study, we showed that high-density octaarginine (R8)-modified liposomes are taken up mainly through macropinocytosis and are delivered to the cytosol while the aqueous phase marker is retained [10]. Therefore, we used R8 as a cytosol delivery device for the MITO-Porter. We also predicted that R8, which mimics TAT, might have mitochondrial-targeting activity [8, 11]. To deliver cargos to target organelles using our strategy, the liposomes must fuse with the organelle membrane. As previously reported, R8-modified envelopes composed of 1,2-dioleoyl-sn-glycero-3-phosphatidylethanolamine (DOPE), and 5-cholesten-3-ol 3-hemisuccinate (CHEMS) showed high transfection activities in dividing cells [12]; however, these lipids may not be the best lipid composition for use in targeting mitochondria.

Therefore, screening was initiated for lipid compositions that are adequate for fusion to the mitochondrial membranes. This was done by monitoring the cancellation of fluorescence resonance energy transfer (FRET) between donor and acceptor fluorophores, modified on the surface of liposomes [13–15]. Liposomal membranes were labeled with both 7-nitrobenz-2-oxa-1,3-diazole (NBD) (excitation at 460 nm and emission at 534 nm) and rhodamine (excitation at 550 nm and emission at 590 nm) so that energy would be transferred from NBD to rhodamine. Membrane fusion between the dual-labeled liposomes and the mitochondria would lead to the diffusion of NBD and rhodamine into the lipid

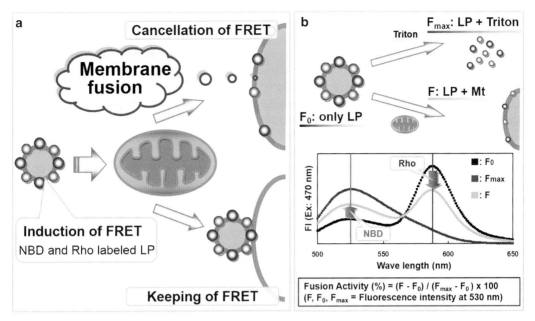

Fig. 2 Schematic image (**a**) and protocol (**b**) for a FRET analysis using isolated mitochondria to evaluate the mitochondrial membrane fusion. *LP* dual-labeled liposome

membranes, which causes a reduction in energy transfer, resulting in an increase in fluorescence intensity at 530 nm (Fig. 2).

To date, we have successfully identified lipid compositions for the MITO-Porter that promote both its fusion with the mitochondrial membrane and the release of its cargo to the intramitochondrial compartment in living cells [7, 16]. In this chapter, we describe the method used for the screening of a series of lipid envelopes that are able to efficiently fuse with isolated rat liver mitochondria. This was achieved by varying the lipid composition of a panel of liposomes and monitoring membrane fusion by FRET analysis. In addition, membrane fusion between the MITO-Porter and mitochondria in living cells was evaluated by FRET analysis using a spectral imaging fluorescent microscopy system [17].

2   Materials

Prepare all solutions using ultrapure water (prepared by purifying deionized water to attain a sensitivity of 18 MΩ cm at 25°C) and commercially available reagent-grade reagents. Prepare and store all reagents at room temperature (unless otherwise indicated).

2.1   Lipids

1. 1,2-dioleoyl-sn-glycero-3-phosphatidylethanolamine (DOPE) (Avanti Polar lipids, Alabaster, AL, USA).

2. 5-cholesten-3-ol 3-hemisuccinate (CHEMS) (Sigma, St. Louis, MO, USA).

3. Cardiolipin (CL) (Sigma).

4. Phosphatidic acid (PA) (Sigma).

5. Phosphatidyl glycerol (PG) (Sigma).

6. Phosphatidyl inositol (PI) (Sigma).

7. Phosphatidyl serine (PS) (Sigma).

8. Sphingomyelin (SM) (Sigma).

9. Cardiolipin (CL) (Sigma).

10. Stearyl octaarginine (STR-R8) (Kurabo Industries Ltd, Osaka, Japan) [18].

11. 7-nitrobenz-2-oxa-1, 3-diazole-labeled DOPE (NBD-DOPE) (Avanti Polar lipids).

12. Rhodamine-labeled DOPE (Rho-DOPE) (Avanti Polar lipids).

2.2 Liposome Preparation

1. Lipid stocks: dissolve 1 mM lipid in 100% ethanol (*see* **Notes 1–4**). Store at –20°C.

2. Mitochondrial isolation buffer (MIB): 250 mM sucrose, 2 mM Tris–HCl, pH 7.4. Store at 4°C.

3. Bath-type sonicator (85 W) (Aiwa Co., Tokyo, Japan).

4. Zetasizer Nano ZS (Malvern Instruments, Herrenberg, Germany).

2.3 FRET Analysis Using Isolated Mitochondria

1. Prepare isolated mitochondria from livers of Wistar rats (6–8 weeks old), as reported by Shinohara et al. [19, 20] (*see* **Notes 5 and 6**). Store on ice.

2. For the FRET analysis, prepare a mitochondrial suspension in MIB (corresponding to the 0.9 mg of mitochondrial protein/mL) (*see* **Note 7**).

3. Triton sol.: 0.5% Triton X-100 (Sigma) in MIB (v/v).

4. Fluorescence spectrometer: FP750 instrument (JASCO, Tokyo, Japan).

2.4 Evaluation of Membrane Fusion in Living Cells Using Spectral Imaging Fluorescent Microscopy

1. HeLa cells: HeLa human cervix carcinoma cells (RIKEN Cell Bank, Tsukuba, Japan).

2. DMEM: Dulbecco's Modified Eagle Medium (Invitrogen Corp., Carlsbad, CA, USA) (*see* **Note 8**).

3. FBS: inactivate fetal bovine serum (Invitrogen Corp.).

4. 35 mm glass-base dishes (IWAKI, Tokyo, Japan).

5. PBS (–): 137 mM NaCl, 2.68 mM KCl, 8.05 mM, Na_2HPO_4, 1.47 mM KH_2PO_4 in water. Autoclave.

6. MitoFluor Red 589 [excitation at 588 nm and emission at 622 nm] (Molecular Probes; Eugene, OR, USA).

7. Confocal laser scanning microscopy (CLSM): LSM510 META (Carl Zeiss Co. Ltd., Jena, Germany).

Table 1
Lipid composition of dual-labeled liposome

1 mM Lipid X[a]	112.5 µL
1 mM Lipid Y[b]	25 µL
0.1 mM NBD-DOPE	13.8 µL
0.1 mM Rho-DOPE	6.9 µL
CHCl$_3$	125 µL

[a]Lipid X is DOPE or EPC
[b]Lipid Y is CL, PI, PG, PS, PA, CHEMS, SM, or Chol

3   Methods

All procedures should be carried out at room temperature, unless otherwise specified.

3.1   Preparation of Dual-Labeled Liposome for the Screening Assay by FRET

1. Add lipid stocks and chloroform to a glass tube following lipid composition list as shown in Table 1 (*see* **Note 9**).

2. Evaporate the organic solvent by means of a vacuum pump to form lipid films on the bottom of the glass tube (*see* **Notes 10** and **11**).

3. Add 0.25 mL of MIB to the dried lipid film on the bottom of a glass tube, and then incubate the suspension for 10–15 min at room temperature to achieve hydration (*see* **Note 12**).

4. Sonicate the suspensions using a bath-type sonicator for 15–30 s to generate empty vesicles (lipid concentration of liposomes, 0.55 mM) (*see* **Note 13**).

5. Measure the particle diameters of the liposomes using a quasielastic light-scattering method, and determine the ζ-potentials by electrophoretically using laser Doppler velocimetry (Zetasizer Nano ZS) (*see* **Notes 14–16**).

3.2   FRET Analysis Using Isolated Mitochondria to Evaluate Fusion with the Mitochondrial Membrane

1. Add a 10-µL aliquot of dual-labeled liposomes to a mitochondrial suspension in 90 µL of MIB to prepare sample [F].

2. As a control without mitochondria, prepare sample [F_0] by adding a 10-µL aliquot of dual-labeled liposomes into 90 µL of MIB.

3. Add a 10-µL aliquot of dual-labeled liposomes into 90 µL of Triton sol. [F_{max}], to define the maximum fluorescence of liposomes dissolved.

4. Incubate these samples for 30 min at 25°C in the dark, after mixing on a vortex mixer.

5. After incubation, measure the fluorescence intensity (excitation at 470 nm and emission at 530 nm) using fluorescence spectrometer to assess the energy transfer (*see* **Note 17**).

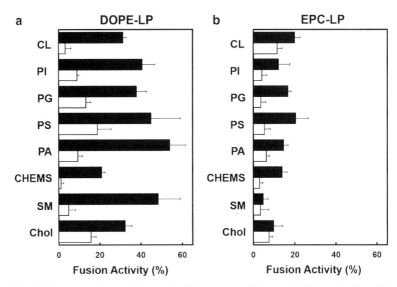

Fig. 3 The screening of fusogenic lipid compositions with the mitochondrial membrane using isolated mitochondria. Fusion activities (%) of DOPE-contained liposomes (**a**) and EPC-contained liposomes (**b**) were calculated in terms of reduction of FRET. Data are presented as the mean of triplicate runs. *Closed bars* R8-LP, *open bars* unmodified liposome [7, 22]

6. Estimate the fusion activity (%) by the reduction in the level of energy transfer as a function of membrane fusion (Fig. 3), which is calculated as follows:

$$Fusion\ activity\,(\%) = \frac{\left(F - F_0\right)}{\left(F_{max} - F_0\right)} \times 100$$

where F, F_0, and F_{max} represent the fluorescence intensity of dual-labeled liposome after incubation with and without mitochondria and the maximum fluorescence intensity after a treatment with Triton X-100, respectively (*see* **Note 18**).

3.3 Evaluation of Membrane Fusion in Living Cells Using Spectral Imaging Fluorescent Microscopy

1. Culture HeLa cells (4×10^4 cells/dish) in 35 mm glass-base dishes with 2 mL of DMEM, which contained 10% FBS, under 5% CO_2/air at 37°C for 24 h.

2. Wash the cells with ice-cold PBS (−), and incubate with the dual-labeled liposomes in 1 mL of serum-free DMEM (final lipid concentration, 5.4 μM) (*see* **Note 19**).

3. Wash the cells with ice-cold PBS (−) after a 1-h incubation under 5% CO_2/air at 37°C, and further incubate the cells in DMEM containing 10% serum for 1 h in the absence of the carriers.

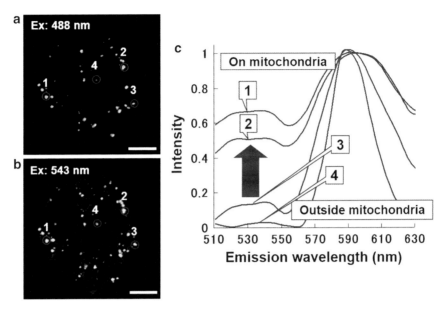

Fig. 4 (**a**) Spectral image figures were observed in the case of excitation at 488 nm to evaluate membrane fusion by FRET. (**b**) Spectral image figures were observed in the case of excitation at 543 nm to identify the localization of liposomes (*yellow color*) and mitochondria (*red color*). (**c**) Fluorescence spectra of ROI were selected from Fig. 4a, and the emission was normalized to the Rho-DOPE peak fluorescence intensity. Scale bars, 10 μm [7]

4. After the incubation, replace with fresh DMEM containing MitoFluor Red 589 (final concentration, 100 nM), and incubate further for 20 min to permit the mitochondria to be stained.

5. After the incubation, wash the cells with ice-cold PBS (–) and then observe them by CLSM (LSM510 META) using an oil-immersion objective lens (Plan-Apochromat 63x/NA = 1.4).

6. To evaluate membrane fusion between the carriers and mitochondria, excite the NBD-DOPE in carriers within cells by light (488 nm) derived from an argon laser. Analyze spectroscopically the emitted light filtered through a dichroic mirror (HFT488) (510–630 nm) as shown in Fig. 4a (*see* **Note 20**).

7. Select fluorescence spectra of the ROI (region of interest) in several signals of Fig. 4a, and normalize the emission at each wavelength by the Rho-DOPE peak fluorescence intensity (emission at 590 nm), as shown in Fig. 4c (*see* **Note 21**).

8. To identify the localization of the carriers on the mitochondria, obtain spectral images ranging from 570 to 630 nm with a He/Ne laser (543 nm) using a dichroic mirror (HFT UV/488/543/633) as shown in Fig. 4b (*see* **Notes 22** and **23**).

4   Notes

1. Dissolve the lipid by sonication using bath-type sonicator, when the lipid is not readily soluble in the solvent.

2. Dissolve the lipid in 100% ethanol/chloroform (1:1, v/v) when PA is used, and store in a glass tube (not a plastic tube).

3. Store lipid stock solutions in tubes wrapped with aluminum foil in the case of fluorescent-labeled lipids (NBD-DOPE and Rho-DOPE).

4. Incubate lipid stocks at room temperature before use.

5. Isolated mitochondria should be freshly prepared each time and should be used as soon as possible.

6. When mitochondria are isolated by this procedure, the purity of the mitochondrial preparation was found to be greater than 90% as judged by electron microscopy observations, and the respiration control index of the mitochondria was determined to be 4.5–6 using a Clark oxygen electrode (YSI 5331; Yellow Springs Instrument Co., Yellow Springs, OH, USA), indicating that the mitochondria were intact and actively respiring. Please *see* refs. 19, 20.

7. Determine the concentrations of mitochondrial proteins using a BCA protein assay kit (Pierce; Rockford, IL, USA), after treatment with a detergent such as sodium dodecyl sulfate to lyse the mitochondria.

8. Use phenol red-free DMEM (Invitrogen Corp.) when the cells are observed.

9. We previously successfully prepared dual-labeled liposomes with various compositions of lipids for screening the fusogenic membrane with mitochondria [7].

10. Mix the lipid solution lightly by vortex mixing and wrap the glass tubes with aluminum foil, before removing the organic solvent by vacuum evaporation.

11. Check that a thin lipid film is formed. If you need to optimize the lipid film, after adding 125 μL of 100% ethanol/chloroform (1:1, v/v) to the glass tubes, remove it by vacuum evaporation.

12. Mix the lipid suspension by vortex mixer before sonication.

13. In the case where the liposomes are modified with R8, add an STR-R8 solution (10 mol% of lipids) to the resulting suspension (liposomes) and then incubate for 30 min at room temperature after gentle mixing.

14. Overall, the ζ-potentials of the neutral lipid-containing liposomes, the anionic lipid-containing liposomes, and the R8-modifeid liposomes should be approximately 0 mV, –50 mV, and +50 mV, respectively [7].

15. The ζ-potentials of the R8-modified liposomes were positively charged, indicating that the surface of the carrier was modified with R8 [7].

16. We find that it is best to prepare the liposomes fresh each time.

17. FRET analysis is frequently used to detect protein interactions in living cells [21]. In this study, we applied this method to evaluate membrane fusion as shown in Fig. 2.

18. As shown in Fig. 3, R8-modified liposomes, which are composed of DOPE/SM/STR-R8 (9:2:1) or DOPE/PA/STR-8 (9:2:1), show a high fusogenic activity with mitochondria.

19. The MITO-Porter used in this experiment is composed of DOPE/SM/STR-R8 (9:2:1, molar ratios). It has been reported that high-density R8-modified liposomes are taken up mainly through macropinocytosis and delivered to the cytosol [10].

20. Through the spectra ranges used in the analysis, background fluorescence derived from the emission from Rho-DOPE and MitoFluor Red 589 was negligible when they were excited at 488 nm (data not shown).

21. In the non-fusogenic state, the emission energy of NBD (donor) is transferred to the excitation energy of rhodamine (acceptor). As a result, the fluorescence intensity of NBD at 530 nm was minimized (lines 3 and 4 in Fig. 4c). Lipid mixing by membrane fusion between the dual-labeled liposomes and mitochondrial membranes results in the diffusion of the donor and acceptor; this results in a reduction in FRET and an increase in the fluorescence intensity at 530 nm (lines 1 and 2 in Fig. 4c).

22. Carriers (yellow) and mitochondria (red) can be clearly distinguished, since the emission spectral peak of MitoFluor Red 589 was detected at the far-red region (emission at 622 nm) compared with that of Rho-DOPE (emission at 590 nm), as shown in Fig. 4b.

23. In Fig. 4c, the spectra present on and outside the mitochondria appear as red or blue lines, respectively. The fluorescence intensity at 530 nm increased significantly, indicating a reduction in energy transfer due to fusion between the liposomal and mitochondrial membranes on the mitochondria. On the other hand, no such increase at 530 nm is evident outside the mitochondria.

Acknowledgements

This work was supported, in part, by the Grant-in-Aid for Young Scientists (A) and Grant-in-Aid for Challenging Exploratory Research from the Ministry of Education, Culture, Sports, Science and Technology of Japanese Government (MEXT). We also thank Dr. Milton Feather for his helpful advice in writing the manuscript.

References

1. Flierl A, Jackson C, Cottrell B, Murdock D, Seibel P, Wallace DC (2003) Targeted delivery of DNA to the mitochondrial compartment via import sequence-conjugated peptide nucleic acid. Mol Ther 7:550–557

2. Vestweber D, Schatz G (1989) DNA-protein conjugates can enter mitochondria via the protein import pathway. Nature 338:170–172

3. Schatz G (1987) 17th Sir Hans Krebs lecture. Signals guiding proteins to their correct locations in mitochondria. Eur J Biochem 165:1–6

4. Seibel M, Bachmann C, Schmiedel J, Wilken N, Wilde F, Reichmann H, Isaya G, Seibel P, Pfeiler D (1999) Processing of artificial peptide-DNA-conjugates by the mitochondrial intermediate peptidase (MIP). Biol Chem 380:961–967

5. Endo T, Nakayama Y, Nakai M (1995) Avidin fusion protein as a tool to generate a stable translocation intermediate spanning the mitochondrial membranes. J Biochem 118:753–759

6. Wiedemann N, Frazier AE, Pfanner N (2004) The protein import machinery of mitochondria. J Biol Chem 279:14473–14476

7. Yamada Y, Akita H, Kamiya H, Kogure K, Yamamoto T, Shinohara Y, Yamashita K, Kobayashi H, Kikuchi H, Harashima H (2008) MITO-Porter: a liposome-based carrier system for delivery of macromolecules into mitochondria via membrane fusion. Biochim Biophys Acta 1778:423–432

8. Yamada Y, Harashima H (2008) Mitochondrial drug delivery systems for macromolecule and their therapeutic application to mitochondrial diseases. Adv Drug Deliv Rev 60:1439–1462

9. Kogure K, Akita H, Yamada Y, Harashima H (2008) Multifunctional envelope-type nano device (MEND) as a non-viral gene delivery system. Adv Drug Deliv Rev 60:559–571

10. Khalil IA, Kogure K, Futaki S, Harashima H (2006) High density of octaarginine stimulates macropinocytosis leading to efficient intracellular trafficking for gene expression. J Biol Chem 281:3544–3551

11. Del Gaizo V, Payne RM (2003) A novel TAT-mitochondrial signal sequence fusion protein is processed, stays in mitochondria, and crosses the placenta. Mol Ther 7:720–730

12. Khalil IA, Kogure K, Futaki S, Hama S, Akita H, Ueno M, Kishida H, Kudoh M, Mishina Y, Kataoka K et al (2007) Octaarginine-modified multifunctional envelope-type nanoparticles for gene delivery. Gene Ther 14:682–689

13. Struck DK, Hoekstra D, Pagano RE (1981) Use of resonance energy transfer to monitor membrane fusion. Biochemistry 20:4093–4099

14. Maier O, Oberle V, Hoekstra D (2002) Fluorescent lipid probes: some properties and applications (a review). Chem Phys Lipids 116:3–18

15. Akita H, Kudo A, Minoura A, Yamaguti M, Khalil IA, Moriguchi R, Masuda T, Danev R, Nagayama K, Kogure K et al (2009) Multi-layered nanoparticles for penetrating the endosome and nuclear membrane via a step-wise membrane fusion process. Biomaterials 30:2940–2949

16. Yamada Y, Akita H, Harashima H (2012) Multifunctional envelope-type nano device (MEND) for organelle targeting via a stepwise membrane fusion process. Methods Enzymol 509:301–326

17. Haraguchi T, Shimi T, Koujin T, Hashiguchi N, Hiraoka Y (2002) Spectral imaging fluorescence microscopy. Genes Cells 7:881–887

18. Futaki S, Ohashi W, Suzuki T, Niwa M, Tanaka S, Ueda K, Harashima H, Sugiura Y (2001) Stearylated arginine-rich peptides: a new class of transfection systems. Bioconjug Chem 12:1005–1011

19. Shinohara Y, Almofti MR, Yamamoto T, Ishida T, Kita F, Kanzaki H, Ohnishi M, Yamashita K, Shimizu S, Terada H (2002) Permeability transition-independent release of mitochondrial cytochrome c induced by valinomycin. Eur J Biochem 269:5224–5230

20. Shinohara Y, Sagawa I, Ichihara J, Yamamoto K, Terao K, Terada H (1997) Source of ATP for hexokinase-catalyzed glucose phosphorylation in tumor cells: dependence on the rate of oxidative phosphorylation relative to that of extramitochondrial ATP generation. Biochim Biophys Acta 1319:319–330

21. Miyawaki A, Llopis J, Heim R, McCaffery JM, Adams JA, Ikura M, Tsien RY (1997) Fluorescent indicators for Ca2+ based on green fluorescent proteins and calmodulin. Nature 388:882–887

22. Yamada Y, Harashima H (2012) Targeting mitochondria: innovation from mitochondrial drug delivery system (DDS) to mitochondrial medicine. Yakugaku Zasshi 132:1111–1118

Chapter 3

Ultrasound-Directed, Site-Specific Gene Delivery

Jason Castle and Steven B. Feinstein

Abstract

With the implementation of gene therapy looming in the near term, an effective delivery system using noninvasive, nonviral-mediated methods appears as an attractive option. This novel platform technology uses gas-filled, ultrasound-directed acoustic microspheres for both diagnostic imaging and therapy and yet may provide a key component for future success in the pursuit of single-gene replacement therapy.

Key words Sonoporation, Ultrasound contrast agents, Microbubbles, Gene therapy

1 Introduction

Ultrasound-directed, site-specific methods for delivering genetic payloads utilize commercial products as carrier vehicles and are safe, efficient, economical, and effective. An ultrasound-based delivery system serves as a platform technology for the treatment of single-gene replacement therapy for monogenomic diseases including a variety of additional ethical drug options. Forthcoming clinical trials designed for gene therapy remain a major goal; although the path to clinical success remains unfulfilled, there is evidence that a successful program will be established in the near future.

On September 4, 2013: Eric Topol, Editor-in-Chief, MedScape, published the following comments regarding gene therapy in MedScape:

> While several years ago it was thought that gene therapy for treatment of diseases would be an abject failure, new evidence has just been published that strongly suggests otherwise. First, noted science journalist Carl had an article in WIRED on the "fall and rise" of gene therapy. Just a week later, 2 articles in Science showed evidence of benefit in 2 rare diseases—one neurodegenerative, one immunodeficiency—using a lentivirus vector, paving the potential for more broad application in the future.

Kewal K. Jain (ed.), *Drug Delivery System*, Methods in Molecular Biology, vol. 1141,
DOI 10.1007/978-1-4939-0363-4_3, © Springer Science+Business Media New York 2014

The delivery of the genetic material (i.e., DNA plasmids, siRNA) requires newer methodologies due, in part, to the limitations associated with viral-mediated therapies. Therefore, the use of gas-filled, acoustic microspheres as delivery vehicles has been promoted. These gas-filled microspheres will be subsequently referred to as microbubbles (MBs) throughout the chapter.

The uses of these MBs have provided clinicians as well as patient diagnostic options for monitoring therapies. When first developed and subsequently approved by the FDA, these diagnostic agents were specifically approved to identify cardiac anatomy. Recently, increased worldwide uses expanded to include whole body imaging. The clinical value of the imaging agents takes advantage of their superior signal-to-noise ratio [1, 2], while the safety and utility in cardiology practices is well established and supported by numerous professional societies [3–5].

Ongoing efforts from researchers around the world expanded the focus of MBs to include novel therapeutic uses including drug delivery methods [6]. This therapeutic platform provides the requisite qualities for widespread clinical utility including safety and efficacy. A brief list of possible applications for the platform technology includes:

1. Nonviral-mediated delivery option for functional nucleic acid (DNA, RNA, siRNA) materials

2. Localized, site-specific delivery of the therapeutic payloads with organ-specific distribution

3. Increased efficiency resulting in an increased therapeutic index associated with reduced systemic toxicity

4. Compounded and novel combination drug/gene therapies

5. New formulations for drugs that formerly exhibited undesirable systemic effects

2   Materials

The following description provides a stepwise procedure for preparation of MBs for gene delivery. Note that many variations of method exist and depend upon the chemistries and physical characteristics that include (a) gas contained within the microsphere, (b) payload to be delivered, (c) target organ/disease, and (d) the hardware and software of the ultrasound system. The fundamental understanding in designing a new therapy must include a variety of seemingly trivial factors that exert a major impact on efficacy.

2.1   Preparation of MBs

Although the concept of MB preparation appears relatively simple, proper mixing whether using commercial agents or noncommercial products is crucial for optimal agent performance and efficiency.

While a gross visualization estimate of the contrast effect within the tissue may appear adequate, improperly mixed MBs can result in marked experimental variation despite a "normal" organ perfusion pattern.

In the clinical setting, an average dose of ultrasound contrast agent may consist of 0.3–0.5 mL which is intravenously injected via a peripheral site. This is the standard approach for achieving left ventricular opacification. Currently, this cardiac application remains the sole FDA-approved indication. Often in the practice of cardiology, as a sonographer is performing standard (non-contrast-enhanced) U/S exam, imaging may not be satisfactory. A MB infusion may be performed to provide better signal-to-nose ratios of the cardiac chambers. A second and even a third injection may be required to fully highlight the cardiac chambers, often based on individual patient variation. These clinical applications and infusion methods are in stark contrast to the precision required in developing preclinical models where standardization of technique is essential.

To activate the MBs, one needs to comply with the manufacturer's specifications included in the package inserts (*see* **Note 1**). All ultrasound contrast agents require a mixing prior to intravascular administration. Certain products utilize commercial mixing equipment, while other products advise brief manual agitation. Following external mixing of the product, the extraction of the vial contents should be performed using an 18–20 gauge needle along with an additional needle for venting. Special attention need be addressed to needle gauge as excessive positive or negative pressure can burst MBs (*see* **Note 2**). Likewise, rapid agent withdrawal may induce exogenous shear forces that can result in MBs disruption and a decreased in MB concentration.

2.2 Mixing of Agent with MB

Prior to performing the intravenous injection of a "mixture" (MBs+ therapeutic payload), the gene/drug should be directly mixed with the MBs within a syringe or object that provides a large surface exposure. In theory there is an ideal ratio of gene/drug to MBs; however, the actual concentrations may be empirically determined based on the outcome data. Here again, a number of factors must be considered to ensure consistency and include the duration of mixing prior to infusion, the physicochemical properties of the agent and payload (electrostatic charge, etc.). The experimental data analysis can be complicated, for example, in the case of performing in vitro analytical measurements of transfection efficiency, one must contend with the physical properties of the MBs; that is, the MBs are somewhat fragile, highly mobile objects that remain buoyant and yet may be disrupted by external shear forces. Our standard procedure accomplishes this external mixing of the gene/drug and MBs within 30 s by inversion of the syringe.

Delivery regimes which rely on co-localization of gene/drug with MBs during ultrasound-directed cavitation in the tissue of interest may require simple mixing. If there is temporal separation of the two solutions, the success of the delivery may be compromised (*see* **Note 3**). The mixing time ensure co-localization may be more important when delivering large molecules such as plasmid DNA or other agents with low bioavailability. In the case of small molecules such as chemotherapeutics, Todorova et al. demonstrated that coadministration may not be essential [7]. The authors augmented metronomic anti-angiogenic chemotherapy with serial injections and cavitation of MBs in tumor vasculature. As a result of cavitation within the tumor vasculature, a reduction of blood flow slowed growth of the tumor. Beyond this, MB cavitation has been shown not only to assist with delivery of agent but also to transiently induce porosity of the lumen of blood vessels. This transient separation of the endothelial cells results in increased drug extravasation thus a higher percentage of the dose reaching the intended target. Likewise, in a recent advancement of the field, researchers in Norway have successfully demonstrated the efficacy of microbubble delivery of chemotherapeutics in a clinical setting. Here they were able to increase treatment tolerance thus prolonging the quality of life in patients with pancreatic adenocarcinoma [8].

Other researchers have proven efficacy of preferentially delivering agent via a MB possessing a charged shell [9, 10]. Cationic MBs have been shown to offer some degree of protection to bound DNA (net negative charge) from DNase degradation [9]. Here plasmid-binding cationic MBs enhanced ultrasound-mediated gene delivery efficiency relative to neutral MBs in both cell culture and mouse hind limb tumors.

Moving one step beyond surface interaction, researchers are capable of incorporating agent into the shell of the MB. Recently developed technology by Artenga Inc (Ottawa, Ontario, Canada) offers an on demand MB production system in which the researcher can specify which targeting biologics are incorporated into the shell [11]. Individual small batch production (~2 mL) capability could be a solution to groups which have difficulty acquiring commercially available microbubbles yet wish to have a fast, reproducible method of generating custom MBs.

During injection, one must be sure to use the proper needle size as well as not pushing too rapidly if a bolus injection is used. If MB/drug is infused, careful attention must be given to ensure proper agitation during injection so that MBs do not float out of solution (*see* **Note 4**). Some lipid-shelled MBs can be infused with standard pumps while the syringe sits on a rocker to prevent separation. Alternatively, there are pumps specifically designed for MB infusion which continuously oscillate throughout the delivery session.

3   Methods

As yet there is no single best method for successful MBs drug delivery. A survey of ten researchers would likely result in ten different recipes. In this chapter, we present the method used to achieve efficacious gene therapy in a preclinical model in our laboratory.

3.1   Delivery of Plasmid DNA for Elevation of High-Density Lipoprotein (HDL)

As our therapeutic goal, we focus on hypoalphalipoproteinemia, and similar dyslipidemias (publication in preparation). In the proposed treatment of these disorders, we deliver plasmid DNA (pDNA) encoding for human apolipoprotein A-I (ApoA-I) (Mirus Bio Corp, Madison, Wisconsin). ApoA-I plays an integral role in the formation of mature HDL [12–14]. A number of recent clinical trials designed to increase total serum HDL via different pathways have not met with success [15, 16] and failed to reach their clinical end points due to a lack of efficacy or unacceptable side effects, and including those using Torcetrapib (Pfizer) and Dalcetrapib (Roche). These technologies have relied upon affecting delayed metabolism or elimination of HDL and resulted in a prolongation of the existing HDL. Our methods differ in that we generate nascent, de novo HDL as the cellular machinery begins producing additional circulating ApoA-I and subsequently HDL. If there is an inherent lack of ability to produce sufficient HDL (i.e., a genetic defect) and its required precursors, simply prolonging an insufficient level or a nonfunctional form of HDL [16] will not generate the desired protective benefit. To date our group has demonstrated therapeutic efficacy in the mouse, rat, and rabbit model.

3.2   Our Method of Gene Therapy in a Rat

Male Sprague–Dawley rats (180–250 g) are anesthetized using inhaled Isoflurane. One researcher prepares the Optison™ (GE Healthcare, Princeton NJ) to ensure proper MB suspension while a second prepares the animal for intravenous delivery of the MBs/ApoA-I plasmids. Prior to imaging the abdomen, all hair should be removed, initially with clippers then with a standard facial razor to ensure adequate acoustic coupling of the transducer (*see* **Note 5**). Using a 26 gauge catheter, the tail vein is cannulated and the output is connected to a catheter with a 3-way stopcock. Once patency is confirmed, the plasmid should be mixed with MBs. In our laboratory, 1 mL of Optison is withdrawn into a 3 mL syringe and 1 mL of the plasmid solution is drawn into the same syringe and mixed by gentle inversion for 30 s. Optison and plasmids are then co-injected to facilitate co-localization of the agent with MBs as ultrasound disruption enables delivery to the target tissue, the liver parenchyma.

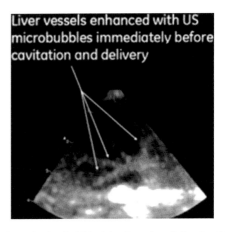

Fig. 1 Upon initiation of microbubble injection via a tail vein of the rat, the contrast effect will be apparent in approximately 20 s in the liver. This perfusion of vasculature is the point at which the sequence of high mechanical index acoustic energy should be initiated to cause MB cavitation and gene delivery. Without this clear visual confirmation efficacy may be compromised

The U/S system is composed of a software-modified Vivid E9 (GE Healthcare Systems, Milwaukee, Wisconsin) permitting customized acoustic pulsing parameters. Location of the liver is initially confirmed by ultrasound examination; and if required, a high-frequency linear array probe (12 L, GE Healthcare Systems) is used until proficiency is established. For MB disruption a lower-frequency cardiac probe (3S, GE Healthcare Systems) is used. Once the injection is initiated, we allow a 15–20 s period of low MI imaging (<0.4 MI) to facilitate the perfusion of the liver parenchyma with the MBs (Fig. 1). During this initial phase, one can identify the presence of the contrast effect. Once liver perfusion has been visually confirmed, the MI is increased (≤1.3 MI) for 2 s before returning to low MI to allow vascular reperfusion. This sequence of alternating high and low MI is repeated for the duration of MB circulation, roughly 2 min in a rat. When the complete dose has been injected, a small bolus of saline is infused to ensure complete administration of dose by flushing the catheter tubing (200 μL in a 0.2 mm line). It is important that a slow rate of injection be used, roughly 30 s/mL to avoid MB disruption and allow adequate opportunity for cavitation and reperfusion during the serial-pulsed sequences. Additionally, a rapid infusion can result in acoustic signal attenuation. In this case, only a limited amount of superficial MBs will be disrupted reducing the delivery of pDNA to the hepatocytes.

Special attention must be given to probe positioning. With the rat placed in a supine position, the ultrasound probe should be placed just caudal to the xiphoid process at approximately a 45° angle directing the beam cranially (*see* **Note 6**). Through empiric studies, we determined that a multi-angle delivery is beneficial.

Naïve Control   Single Plane   Multi Plane

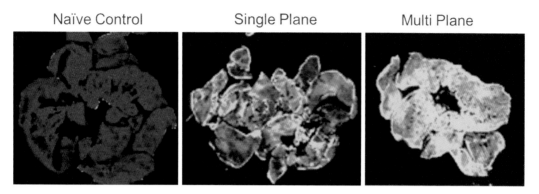

Fig. 2 For early proof of concept studies pDNA coding for green fluorescent protein was used to verify efficacy of microbubble gene delivery to the liver of naïve rats. Additionally, demonstrated is the effect of single versus multiple planes of insonation during gene therapy. As can be seen above, a higher degree of reporter gene expression is achieved when directing U/S energy in multiple planes

That is, following each 2 s period of high MI, the probe should be moved slightly and the angle of insonation changed to provide U/S energy to each of the lobes of the liver over the course of the 2-min delivery session. This should include increasing and decreasing the angle as well as directing acoustic energy laterally. Given the size of the rat, and using an unfocused ultrasound beam centered at 2 cm below the skin, a relatively large portion of the liver is captured in each two-dimensional slice of acoustic pulsing. This approach is the reason that probe angulation and positioning is important in maximally interrogating all lobes of the liver (Fig. 2). If it is difficult to achieve sufficient contrast effect within the liver parenchyma, improper probe placement is likely.

Note that while using a variable angulation of the probe is beneficial, care must be given to limit off-target delivery of the agents resulting in decreased efficacy. The liver focus can be maintained by using direct visualization of liver perfusion. Additionally, minimal external mechanical pressure should be applied to the probe as excessive pressure may reduce liver perfusion.

For our studies, a variety of parameters were used to monitor animal health including a veterinary diagnostic blood analyzer (VetScan VS2, Abraxis Inc, Union City, California) that assesses 14 parameters of blood (albumin, alkaline phosphatase, alanine aminotransferase, amylase, blood urea nitrogen, calcium, creatinine, globulin, glucose, potassium, sodium, phosphorus, total bilirubin, and total protein). No significant deviation from baseline has been observed. Additionally, gross histological analysis of liver following treatment revealed no adverse morphological changes.

While numerous associated experimental components can and need to be addressed and controlled during the procedures to achieve an optimal outcome, our data support the concept that additional chemical binding of agents to the MB is not required.

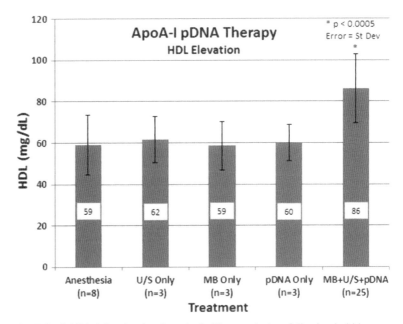

Fig. 3 Peak HDL following treatment of either control or full microbubble gene therapy. The combination of co-injection of microbubble and ApoA-I pDNA with ultrasound acoustic energy (MB + U/S + pDNA) directed at the liver resulted in a significant elevation of serum HDL (mg/dL). This HDL increase followed expression of human mRNA, resulting in production of human ApoA-I protein in the rat within 12 h of treatment. This effect was not seen in any of the control cohorts. Average HDL value (mg/dL) for each cohort is presented on the respective bar

Our results are specific for an agent that has low bioavailability (rapid sequestration/metabolism in systemic circulation) and poor native efficacy.

As researchers undertake a program of microbubble gene therapy, they should bear in mind that while none of the steps or procedures stated above is complex, it is a magnification of minor errors that can result in loss of experimental efficiency. Seemingly trivial details become manifest and demand that an appropriate degree of attention is provided in a consistent manner. To this end by performing our studies with meticulous attention to detail, we achieved 93 % study efficacy. Figure 3 is a representative data set that is a result of the materials and methods described above. These results are noteworthy due to the inherent variability associated with performance of in vivo systems.

4   Notes

1. To activate microbubbles, one must comply with the manufacturer's specifications included in the package inserts. Alternatively, follow methods presented in literature if preparing a unique MB on the bench top.

2. Special attention need be addressed to needle gauge as excessive positive or negative pressure can burst MBs. Using improper gauge needles or drawing too rapidly can reduce the concentration of MBs thereby negatively effecting delivery.

3. Delivery regimes which rely on co-localization of gene/drug with MBs during ultrasound-directed cavitation in the tissue of interest may require simple mixing. If there is temporal separation of the two solutions, the success of the delivery may be compromised. Mixing times will be dependent on bubble and drug properties. More complicated techniques in which agent is bound to or incorporated into the MB shell will require more elaborate procedures.

4. Whether bolus or a longer duration infusion, the proper rate of injection must be used to maximize drug delivery. Pushing too rapidly will allow a significant percentage of drug/MB to pass through the target tissue without the benefit of sonoporation. Conversely, for longer injections, careful attention must be given to ensure proper agitation so that MBs do not float out of solution.

5. Prior to imaging the abdomen, all hair should be removed, initially with clippers then with a standard facial razor to ensure adequate acoustic coupling of the transducer. It is very important to ensure as much fur as possible to prevent acoustic interference. One technique is to shave as stated above, apply acoustic coupling gel to the area. This prevents the skin from drying out while the catheter is being placed.

6. Special attention must be given to probe positioning. With the rat placed in a supine position, the U/S probe should be placed just caudal to the xiphoid process at approximately a 45° angle directing the beam cranially. Through numerous studies to identify the best probe mechanics, it was demonstrated that a multi-angle delivery is required. Using a 2D system treats only a single plain in each therapeutic pulse. To reach as much target tissue as possible (e.g., liver) each pulse should be aimed in a slightly different direction.

The experimental protocol of this study conformed to the Guide for the Care and Use of Laboratory Animals and was approved by the IACUC at the General Electric Global Research (Niskayuna, NY), which is an AAALAC-, USDA-, and OLAW-accredited facility.

Acknowledgements

The authors would like to thank all members of the team who have made microbubble gene delivery a success in our lab. This includes physicist Kirk Wallace, ultrasound engineer David Mills, Chemists

Matthew Butts, Bruce Johnson, and Binil Kandapallil, biologists Mike Marino, Chris Morton, Jeannette Roberts and Andrew Torres as well as the leadership team who make it possible.

The project described was supported in part under NIH SBIR 1R44HL095238 "Development of Novel Tissue Directed Ultrasound Therapeutic Gene Delivery System" from DHHS, National Heart, Lung and Blood Institute. The content is solely the responsibility of the authors and does not necessarily represent the official views of DHHS, NIH, National Heart, Lung and Blood Institute as well as funds from GE Global Research, Niskayuna NY.

References

1. Feinstein SB (2004) The powerful microbubble: from bench to bedside, from intravascular indicator to therapeutic delivery system, and beyond. Am J Physiol Heart Circ Physiol 287(2):H450–H457

2. Kurt M, Shaikh KA, Peterson L et al (2009) Impact of contrast echocardiography on evaluation of ventricular function and clinical management in a large prospective cohort. J Am Coll Cardiol 53(9):802–810

3. Mulvagh SL, Rakowski H, Vannan MA et al (2008) American Society of Echocardiography consensus statement on the clinical applications of ultrasonic contrast agents in echocardiography. J Am Soc Echocardiogr 21(11): 1179–1201

4. Senior R, Becher H, Monaghan M et al (2009) Contrast echocardiography: evidence-based recommendations by European Association of Echocardiography. Eur J Echocardiogr 10(2): 194–212

5. Intersocietal Accreditation Commission (2010) IAC standards and guidelines for adult echocardiography accreditation. http://www.intersocietal.org/echo/standards/IACAdultEchocardiographyStandards2012.pdf. Accessed 07 Sept 2013

6. Castle J, Butts M, Healey A et al (2013) Ultrasound medicated targeted drug delivery; recent success and remaining challenges. Am J Physiol Heart Circ Physiol 304:H350–H357

7. Todorova M, Agache V, Mortazavi O et al (2013) Antitumor effects of combining metronomic chemotherapy with the antivascular action of ultrasound stimulated microbubbles. Int J Cancer 132(12):2956–2966

8. Kotopoulis S, Dimcevski G, Gilja OH et al (2013) Treatment of human pancreatic cancer using combined ultrasound, microbubbles, and gemcitabine: a clinical case study. Med Phys 40:072902

9. Sun L, Huang CW, Wu J et al (2013) The use of cationic microbubbles to improve ultrasound-targeted gene delivery to the ischemic myocardium. Biomaterials 34(8):2107–2116

10. David S, Wang MD, Panje C et al (2012) Cationic versus neutral microbubbles for ultrasound-mediated gene delivery in cancer. Radiology 264:721–732

11. Goertz DE, Todorova M, Mortazavi O et al (2012) Antitumor effects of combining docetaxel (taxotere) with the antivascular action of ultrasound stimulated microbubbles. PLoS One 7(12):e52307. doi:10.1371/journal.pone.0052307

12. Zannis VI, Chroni A, Krieger M (2006) Role of apoA-I, ABCA1, LCAT, and SR-BI in the biogenesis of HDL. J Mol Med (Berl) 84(4):276–294

13. Davidson MH (2010) Update on CETP inhibition. J Clin Lipidol 4:394–398

14. Duffy D, Rader DJ (2009) Update on strategies to increase HDL quantity and function. Nat Rev Cardiol 6:455–463

15. Nissen SE, Tsunoda T, Tuzcu EM et al (2003) Effect of recombinant ApoA-1 Milano on coronary atherosclerosis in patients with acute coronary syndromes. JAMA 290(17):2292–2300

16. Yamamoto S, Yancey PG, Ikizler TA et al (2012) Dysfunctional high-density lipoprotein in patients on chronic hemodialysis. J Am Coll Cardiol 60(23):2372–2379

Chapter 4

Synthesis of Thermoresponsive Polymers for Drug Delivery

Sushil Mishra, Arnab De, and Subho Mozumdar

Abstract

A protocol for synthesizing thermosensitive copolymers of N-isopropylacrylamide (NIPAM) and N-vinylpyrrolidone (VP), cross-linked with N,N′-methylene-bis-acrylamide (MBA) has been described in this chapter. The copolymers have been formed at different concentrations of NIPAM and VP and at two different temperatures (70 °C and 30 °C). The lower critical solution temperature (LCST) of the samples has been measured, and the size of the particles formed with the highest concentration of NIPAM and lowest concentration of VP (MG1 and NG1) has been measured at three different temperatures of 25 °C, 35 °C, and 37 °C. Both MG1 and NG1 showed the lowest size at 37 °C. The MG1 and NG1 samples were further characterized using TEM and SEM. The MG1 particles were subsequently used for protein drug delivery, using BSA as a model. The release profile showed the best fit with the zero-order model. Finally, cytotoxicity studies of the synthesized MG1 and NG1 particles were carried out, using in vitro MTT assay, so as to determine the overall biocompatibility of the materials.

Key words Thermoresponsive, N-Isopropylacrylamide, Microgel, Nanogel, Drug delivery system, Polymer cytotoxicity

---

1 Introduction

A variety of biological processes are controlled by feedback-mechanism in the body. The body can communicate through a chemical language which involves nucleic acids, proteins, and polypeptides (that have the ability to adopt conformations specific to their surroundings), about own demand to run the system and also upon fulfillment stop the process. Similar behavior has inspired the development of biomimetic strategies in both natural and synthetic polymers for a series of applications, including in bioactive delivery [1, 2].

The polymers have been used traditionally for protection of drug molecules while delivering through stressful conditions. We can talk about synthetic (co)polymers whose utility can go beyond providing structural support and instead allow active participation in a dynamic sense. These polymers can have numerous copies of functional groups that can be amenable to a change in character

Kewal K. Jain (ed.), *Drug Delivery System*, Methods in Molecular Biology, vol. 1141,
DOI 10.1007/978-1-4939-0363-4_4, © Springer Science+Business Media New York 2014

and thus can bring about synergistically amplified dramatic transformations in the macroscopic material properties [1, 2].

Target-specific drug delivery controlled by physiochemical stimulus of the body offers a significant challenge [3]. Advances in polymer science for biological and medical research have led to the development of several novel intelligent polymeric materials for controlled drug-delivery systems. These materials are capable of self-regulating the delivery of a bioactive agent in response to a specific stimulus. In this direction, new delivery systems are being developed so as to deliver a therapeutic compound to the intended site of action at a predetermined rate and also to maintain the drug concentration in the therapeutic window. The delivery agents for this purpose are mostly polymeric carriers which can safeguard the drug till it reaches the site of action. Furthermore, they can regulate the release of drug so as to obtain a desired release profile. These delivery systems can be designed to act smartly and respond to some physiological or biological stimuli which can act as a language for them and supervise as how and when to work and when to stop [4, 5].

Biodegradable polymers find widespread use in drug delivery as they can be potentially degraded to nontoxic materials inside the body [6]. The routine metabolic systems of the body are well equipped to slowly remove them, thus preventing unnecessary accumulation.

Smart polymers are the materials composed of monomers that can respond in a dramatic way to very slight changes in their environment [7]. Smart polymers are becoming increasingly more common as scientists are learning about their chemistry and triggers that can activate conformational changes in the polymer structures [8].

This chapter is contains three methods: (1) design, synthesis, and characterization of a thermoresponsive polymer; (2) the application of a thermoresponsive polymer for drug delivery by taking bovine serum albumin (BSA) as drug model; and (3) evaluation of cytotoxicity of the polymer.

1.1 Design of Thermoresponsive Drug Delivery Systems

A proper consideration of surface and bulk properties can help in the designing of polymers for various drug delivery applications [6]. New polymeric materials are being chemically formulated that can sense specific environmental changes in biological systems and can adjust in a predictable manner, making them a useful tool for drug delivery [8].

Smart polymers can be designed to respond to different stimuli including light [9, 10], temperature [4, 11–13], pH [14–17], electric fields [18–20], ultrasonic energy [21, 22], or response to chemicals such as glucose [23–26], enzymes [27–29], antigen [30–32], redox/thiol [33, 34], etc.

Many attempts have been made in recent years for developing formulations leading to the control release of chemicals (especially in drug delivery systems). Control of adsorption, permeation, and enzyme reaction using stimuli-responsive polymer gels have also been attempted in the area of chemical and bioengineering. The selection and design of a polymer can be a challenging task because of the inherent diversity of structures and can require a comprehensive perception of the surface and bulk properties of the polymer so as to generate the preferred chemical, interfacial, mechanical, and biological functions.

1.2 Delivery of Thermoresponsive Polymers

Thermoresponsive polymers can undergo major changes in conformation in response to slight changes in the temperature. The temperature can be one of the most widely used stimulus for triggering a signal and this is borne by the fact that the body temperature can often deviate from the physiological value of 37 °C in the presence of pathogens or pyrogens. The change of temperature is relatively easy to control and can be easily applied under both in vitro and in vivo conditions. The deviation in the temperature can be utilized to stimulate the active release of therapeutic agents from various temperature-responsive drug delivery systems for diseases accompanied by fever [35–38].

The thermoresponsive polymers possess a unique property and that is the presence of a critical solution temperature. Critical solution temperature is the temperature at which the polymer undergoes phase transition (volume phase-transition or sol–gel phase-transition) and solution is discontinuously changed according to their composition. The thermoresponsive polymers can be categorized into two types—the polymers which are initially soluble but become insoluble upon heating and thus have lower critical system temperature (LCST) and the other type where the polymer remains initially insoluble but becomes soluble upon heating and thus have upper critical system temperature (UCST) (*see* Fig. 1) [35–38].

LCST polymers include poly(N-isopropylacrylamide), poly(N,N-diethylacrylamide), poly(N-ethylmethacrylamide), poly(methyl vinyl ether), poly(2-ethoxyethyl vinyl ether), poly(N-vinylisobutyramide), poly(N-vinyl-n-butyramide), poly(N-vinylcaprolactam), polyphosphazene derivatives, and poly(N-(2-hydroxypropyl) methacrylamide mono/di-lactate). UCST polymers include networks of poly(acrylic acid) and polyacrylamide or poly(acrylamide-co-butylmethacrylate), poly(ethylene oxide)-b-poly(propylene oxide)-*b*-poly(ethylene oxide) (commercially known as Pluronics, Poloxamer) [36].

N-Isopropylacrylamide (NIPAM) is a temperature-responsive monomer that was first synthesized in the 1950s [39]. It forms a three-dimensional hydrogel when cross-linked with N,N′-methylene-bis-acrylamide (MBA) or N,N′-cystamine-bisacrylamide. When heated in water above 33 °C, it undergoes a reversible LCST

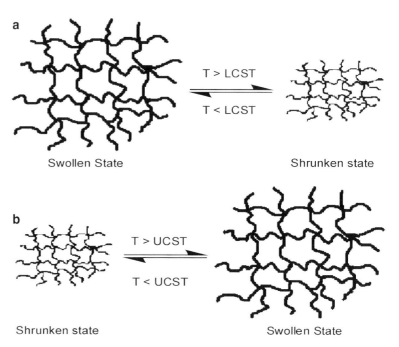

Fig. 1 Shows thermoresponsive behavior of polymeric/copolymeric system

phase transition from a swollen hydrated state to a shrunken dehydrated state, losing about 90 % of its mass. Since poly-NIPAM (PNIPAM) expels its liquid contents at a temperature near that of the human body, PNIPAM has been investigated by many researchers for possible applications in controlled drug delivery [11, 40].

A thermoresponsive copolymeric system based on N-isopropylacrylamide, acrylic acid N-hydroxysuccinimide, acrylic acid, and acryloyloxy dimethyl-c-butyrolactone has been synthesized. The copolymers underwent a rapid sol–gel transition and formed gel quickly following injection so as to form a solid drug depot in situ while preventing mass efflux of free drug into the surrounding environment [41].

Karir et al. have demonstrated the application thermosensitive polymer based on poly N-isopropylacrylamide and L-tyrosinamide for locoregional radionuclide therapy [42]. Their copolymer system underwent phase separation at 35 °C. The in vivo studies in mice showed higher retention of the radiolabeled polymer at the tumor site when injected intratumorally. The radiotherapy targeting of the tumor site would not only destroy the necrotic cells but also cause less harm to the rest of the healthy cells of the body.

The Hexamethylene diisocyanate has been introduced into Pluronic F127 as a chain extender and then incorporated with hyaluronic acid to develop a thermoresponsive nanocomposite hydrogel system. The results showed that nanocomposite hydrogel system could undergo sol–gel transition as temperature was increased to 37 °C. The nanocomposite polymer could spontaneously self-assemble into micellar structure with size of 100–200 nm. Moreover, the release of doxorubicin from the composite hydrogel system showed a zero-order profile and maintained sustained release for over 28 days. The cell viability test on MCF-7 human breast tumor cells showed significant decrease of tumor cell viability. The in vivo mice study demonstrated antitumor efficacy for 4 weeks indicating the potential to have therapeutic effect for cancer therapy [43].

A cutaneous wound dressing hydrogel system containing poly-hexamethylene biguanide has been developed by incorporating acryoyl-lysine into poly(ethylene glycol) cross-linked PNIPAM. At room temperature, the hydrogel remained highly swollen and translucent, but at body temperature the same hydrogel collapsed and became opaque. The application of the hydrogels to a rodent cutaneous wound healing model resulted in significant increase in healing rate when compared with controls. Moreover, the hydrogels were also able to decrease bacteria levels in an infected wound model [44].

Tan el al. [45] have developed an injectable hydrogel for adipose tissue engineering. A thermosensitive copolymer hydrogel with aminated hyaluronic acid-g-poly(N-isopropylacrylamide) (AHA-g-PNIPAAm), was synthesized by coupling carboxylic end-capped PNIPAM to AHA through amide bond linkages. The LCST of AHA-g-PNIPAAm copolymers in phosphate buffer saline was nearly 30 °C. Encapsulation of human adipose-derived stem cells (ASCs) within hydrogels showed that the AHA-g-PNIPAAm copolymers were non-cytotoxic and preserved the viability of the entrapped cells.

This chapter describes synthesis of a thermoresponsive copolymer of N-isopropylacrylamide (NIPAM) and N-Vinylpyrrolidone (VP). The N-Vinylpyrrolidone (VP) has been attractive in the chemical, pharmaceutical, and material fields because of the combination of properties including enhanced solubility in water and organic solvents, nontoxic nature, biocompatibility, and high complexation ability [46–49]. Copolymerization is a valuable method for tailoring a polymeric product so as to have specific properties. Copolymerization modifies the symmetry of the polymer chain and modulates both intermolecular and intramolecular forces, so that the properties (such as LCST, cloud point, solubility, and so on) of the smart polymer can be appropriately tuned.

1.3 Evaluation of Cytotoxicity of a Polymer

Polymers are potentially cytotoxic. Evaluation of cytotoxicity is important because the polymer to be used in drug delivery needs to be clinically safe. The polymer also needs to be biocompatible. Moreover, demonstration of safety is important in demonstrating safety in commercial pharmaceutical setting.

2   Part A: Synthesis of Thermoresponsive Polymer

2.1   *Material*

1. N-isopropylacrylamide (*see* Fig. 2a) is procured from Acros Organics, Belgium. Further, it is recrystallized from n-hexane and stored at 4 °C before use.

2. Vinyl pyrrolidone (*see* Fig. 2b) is obtained from Lancaster, USA and freshly distilled before polymerization.

3. N, N'-methylene-bis-acrylamide (*see* Fig. 2c) is obtained from Sigma Chemicals (St. Louis, MO, USA). N,N'-methylene-bis-acrylamide solution of 0.049 g/mL is prepared in double-distilled water.

4. Ammonium persulfate is procured from SRL (Mumbai, India). A saturated aqueous solution of ammonium persulfate is freshly prepared before use.

5. The dialysis membrane of MWCO 12 kDa is procured form Sigma. The dialysis membrane is soaked overnight and stored at 4 °C. Dialysis membrane is rinsed thoroughly with double-distilled water for several times before use.

6. Double-distilled water is used throughout the experiment.

2.2   *Method*

1. Take pre-weighed quantity of N-isopropylacrylamide and vinylpyrrolidone in a round bottom flask and add 10 mL of double-distilled water.

2. The system is maintained in constant stirring in a water bath at a temperature of 70 °C for microgel particle synthesis and at 30 °C for nanogel particle synthesis, respectively. The molar ratios of these two monomers are varied according to the tabulated data (*see* Table 1).

3. At an interval of half an hour, the next addition of reaction material MBA (30 μL) and ammonium persulfate (5 μL) are made in the above reaction mixture.

4. Following the final additions, the reaction mixture is allowed to stir under nitrogen atmosphere for 5 h for microgel particle synthesis and 24 h for nanogel particle synthesis at the respective temperatures (*see* Fig. 3).

5. The obtained copolymer is dialyzed in double-distilled water for 72 h using a dialysis membrane. The dialysis medium is

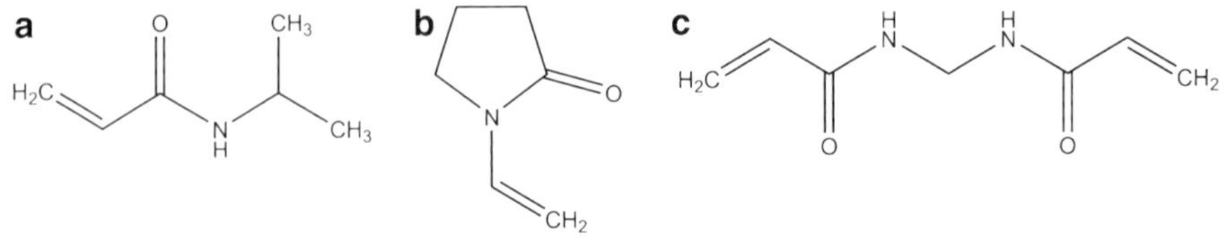

Fig. 2 Chemical structure of (**a**) of N-isopropylacrylamide, (**b**) Vinylpyrrolidone, and (**c**) N,N'-methylene-bis-acrylamide

Table 1
The table shows the various molar ratios of NIPAM and VP altered to synthesize the different copolymer samples for microgel (MG) and nanogel (NG)

Composition	MG/NG-1	MG/NG-2	MG/NG-3	MG/NG-4	MG/NG-5
NIPAM	9	7	5	3	1
VP	1	3	5	7	9

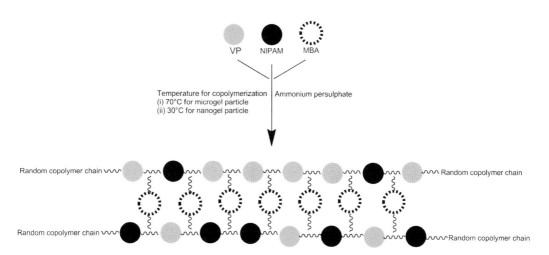

Fig. 3 Flow diagram for synthesis of copolymeric particles form NIPAM and VP

replaced with fresh double-distilled water at every 4 h within first 24 h and at 6 h for 48–72 h.

6. The dialyzed copolymer solution is frozen by keeping the round bottom flask in liquid nitrogen.

7. Further, the water is completely removed from the copolymer solution by lyophilizing the frozen samples (*see* **Note 1**).

2.3 Cloud Point Measurement (Phase Transition Temperature Analysis)

Cloud point is determined as the temperature at which the first sign of turbidity appears in the solution. Below the cloud point temperature, the aqueous copolymer solution appears clear but upon heating the solution becomes turbid because of possible aggregation of the polymers. The cloud point in this experiment is referred as lower critical solution temperature (LCST) of the polymer solution.

1. The cloud point temperature is recorded in a thermostable water bath with the aqueous solution of copolymers synthesized (*see* Subheading 2.2).

2. The copolymers are weighed and taken in sealed glass tubes then shaken to disperse them uniformly.

3. Absence of leakage is ensured by weighing the glass tubes before and after the cloud point measurements.

4. The temperature is raised gradually with an increase of 0.5 °C/min.

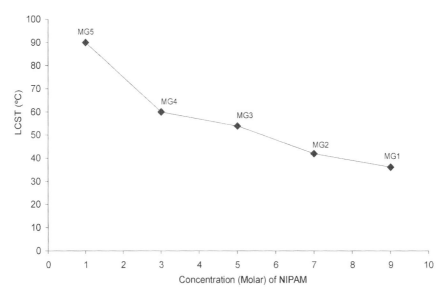

Fig. 4 LCST analysis of the microgel particles MG1 (37 °C), MG2 (42 °C), MG3 (54 °C), MG4 (60 °C), and MG5 (No phase transition observed till 80 °C). MG1 shows LSCT at human physiological temperature

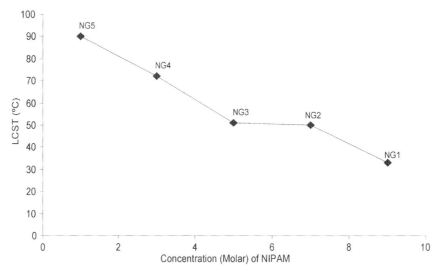

Fig. 5 LCST analysis of the nanogel particles NG1 (34 °C), NG2 (50 °C), NG3 (51 °C), NG4 (72 °C), and NG5 (No phase transition observed till 80 °C). NG1 shows LSCT near to human physiological temperature

5. Cloud point is determined visually at the first sign of appearance of turbidity in the solution with a thermometer having an accuracy of ±0.2 °C. This cloud point is termed as LCST of the copolymer sample solution.

6. The reproducibility of the result is checked within ±0.4 °C.

7. A graphical representation is established for LCST versus molar concentration of NIPAM in the copolymers for microgel and nanogel particle synthesis (*see* Figs. 4 and 5).

2.4 Quasi Elastic Light Scattering Analysis

The Quasi Elastic Light Scattering (QELS) instrument (Photocor FC) is used for analyzing the size of the synthesized thermoresponsive particles (*see* **Note 2**).

1. Weigh 10 mg of lyophilized thermoresponsive particles in 10 mL double-distilled water (*see* Subheading 2.2).

2. Sonicate the samples and incubate them at the pre-mentioned temperature for 24 h in advance.

3. The particle size analysis for microgel particles are carried out at three different temperatures, viz., 25 °C, 35 °C, and 37 °C.

4. The particle size analysis for nanogel particles are carried out at three different temperatures, viz., 25 °C, 34 °C, and 37 °C.

5. The particle size and polydispersity index (PDI) are measure by the DYNALS software (*see* Fig. 6).

2.5 Electron Microscopic Analysis of the Synthesized Particles

1. In order to obtain scanning electron microscopy (SEM) images of the samples (*see* Fig. 7), a colloidal drop of suspension is dropped an aluminum stub (*see* **Note 3**).

2. The SEM samples are dried at ambient temperature and then 5 nm thick gold layer is spread on the sample surface before imaging.

3. In order to obtain transmission electron microscopy (TEM) images of the samples (*see* Figs. 8 and 9), a colloidal drop of suspension is dropped on a carbon coated grid and dried further (*see* **Note 4**).

3   Part B: In Vitro Release Kinetics Models for Microgel Particle (MG1)

3.1 Material

1. Phosphate Buffer (pH of 7.4) is prepared by placing 50.0 mL of the 0.2 M potassium dihydrogen phosphate in a 200-mL volumetric flask, add 39.1 mL of 0.2 M sodium hydroxide and then add double-distilled water to make up the volume (*see* **Note 5**) [50].

2. Bovine serum albumin is procured from CDH (India) and stored at 4 °C.

3. Dialysis membrane (*see* Subheading 2.1, **item 5**).

3.2 Method

3.2.1 Entrapment of Bovine Serum Albumin (BSA) in NIPAM-VP Microgel Particles (MG1)

1. Weighed amounts of the microgel particles (10 mg) and BSA (10 mg) are added to double-distilled water.

2. The model protein BSA is incubated with the microgel particles (MG1) for 4 days at 4 °C to attain the swelling equilibrium.

3. The loaded microgel particles are dialyzed in double-distilled water for 72 h using a dialysis membrane for the removal of

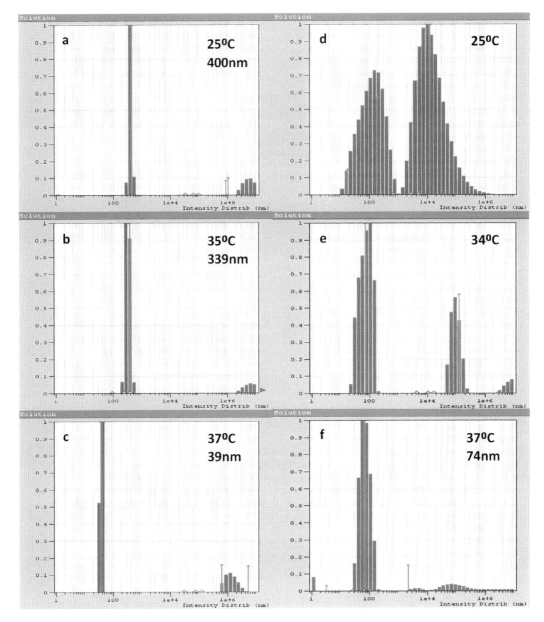

Fig. 6 QELS picture of the microgel (**a–c**) and nanogel (**d–f**) copolymer at various temperatures. In both the cases minimum size is observed at 37 °C

free BSA in the dispersion or that adsorbed on the surface of the microgel particles. The dialysis medium is replaced with fresh double-distilled water at every 4 h within first 24 h and at 6 h for 48–72 h.

4. The dialyzed material is taken in a round bottom flask and frozen by keeping it in the liquid nitrogen.

5. The frozen microgel particles are lyophilized to dry powder using a freeze drier (*see* **Note 1**).

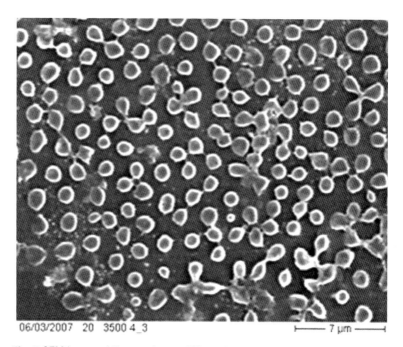

06/03/2007   20   3500 4_3   ├──── 7 μm ────┤

Fig. 7 SEM image of the copolymer (MG1) microgel at 7 μm resolution

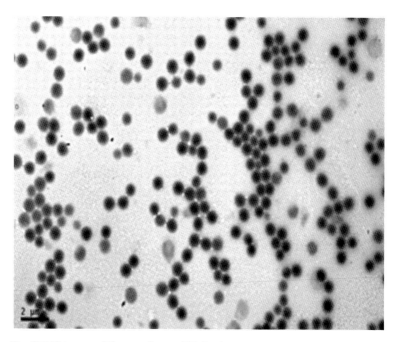

Fig. 8 TEM image of the copolymer (MG1) microgel at 2 μm resolution

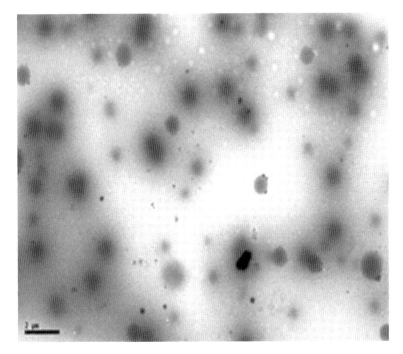

Fig. 9 TEM image of the copolymer NG1 nanogel at 2 μm resolution

3.2.2 Entrapment Efficiency of BSA Loaded Microgel Particle

1. An accurately weighed quantity of 10 mg of lyophilized formulation containing BSA is dispersed in 50 mL of phosphate buffer (pH of 7.4).

2. The extraction of entrapped BSA in the lyophilized product is done by sonicating the samples for 1 h with intermitted shaking followed by centrifugation at 10,000 rpm ($12857 \times g$) for 10 min.

3. Filter the solution (*see* **Note 6**) by discarding the initial few drops of the sample.

4. Measure the absorbance of BSA at 278 nm against the plain buffer (pH of 7.4) setting as a blank (*see* **Note 7**). The reading is recorded by using a UV–Vis spectrophotometer (*see* **Note 8**).

5. The entrapment efficiency ($E\%$) is calculated by using Eq. 1:

$$Entrapment\ Efficiency\left(E\%\right) = \frac{Mass\ of\ drug\ in\ microparticle}{Mass\ of\ drug\ used\ in\ formulation} \times 100 \qquad (1)$$

3.2.3 Release Kinetics Study of BSA

The in vitro release kinetics of the thermoresponsive delivery carrier is determined for their capability of drug release property, in which BSA is used as a model protein for the release study (*see* Fig. 10).

1. 40 mL of freshly prepared phosphate buffer at pH 7.4 is used as the dissolution medium.

2. The temperature of 37 ± 0.5 °C is maintained throughout the release study.

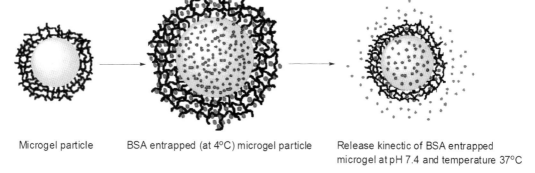

Microgel particle   BSA entrapped (at 4°C) microgel particle   Release kinectic of BSA entrapped
                                                                            microgel at pH 7.4 and temperature 37°C

Fig. 10 Schematic representation of the encapsulation of BSA and temperature-dependent release form microgel particle (MG1)

3. The samples are taken at the time points of 5 min, 10 min, 20 min, 30 min, 40 min, 60 min, 75 min, 90 min, and 120 min.

4. The drug release is evaluated by withdrawing 3 mL of the sample and replacing it with an equal quantity of fresh medium.

5. The samples are filtered (*see* **Note 6**) before taking UV-absorbance.

6. The amount of BSA released at each time interval from the microgel particles into the buffer solution is measured by absorbance of BSA at 278 nm against the plain buffer (at pH of 7.4), setting as a blank (*see* **Note 7**). The reading is recorded by using a UV–Vis spectrophotometer (*see* **Note 8**).

7. The following plots are established over the obtained results for the BSA release profiles (*see* Fig. 11):

   (a) Cumulative % drug release vs. time (Zero-order kinetic model)

   (b) Log cumulative of % drug remaining vs. time (First order kinetic model)

   (c) Cumulative % drug release vs. square root of time (Higuchi model)

   (d) Cube root of drug % remaining in matrix vs. time (Hixson–Crowell cube root law).

4   Part C: In Vitro Cytotoxicity Assessment of the Thermoresponsive Copolymers

All compounds have the potential to be poisonous depending upon dose. This observation makes toxicity testing an essential and critical parameter for industrial practice to identify and define safety thresholds for all compounds for biological use. The in vitro cytotoxicity tests is based on the principle that toxic chemicals affect the basic functioning of the cells (common to all cells) and the level of toxicity can be measured by assessing cellular damage [51].

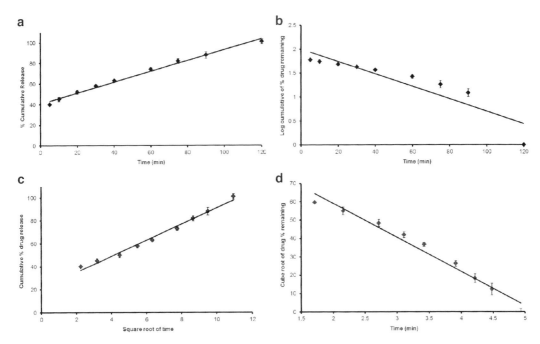

Fig. 11 Shows various kinetic release models for BSA as drug form microgel particles; (**a**) Zero-order kinetic model, (**b**) First order kinetic model, (**c**) Higuchi model, and (**d**) Hixson–Crowell cube root law. The best linearity is shown by zero-order equation plot where r^2 is 0.9909

There are three types of cytotoxicity assays which are performed to assess the cytotoxicity of any material. These cytotoxicity assays are (a) membrane integrity/viability, (b) cellular functioning, and (c) cell viability assays [52–54]. Membrane integrity assays assess the cell damaged when they become leaky [53]. Membrane integrity is determined by measuring lactate dehyrogenase (LDH) [55, 56] in the extracellular medium. Cellular functioning assay measures the cellular metabolic activity caused by cell damage. Tests which can measure metabolic function measure cellular ATP levels or mitochondrial activity (via MTT/MTS metabolism) [54]. The cell viability assay [57] is the direct measure of cell number, since dead cells normally detach from a culture plate and are washed away in the medium. Cell number can be measured by direct cell counting or by the measurement of total cell protein or DNA, which are proportional to the number of cells.

The ability to measure the early indications of toxicity of both the drug and the drug carrier is an essential component in the development of any drug delivery vehicle. Therefore, in order to confirm the lack of any toxic properties (predominantly at the early stages of the development) of a drug carrier, all the above mentioned cytotoxicity assays have been extensively measured since long. Libraries of newly synthesized polymers with an intended use in the field of drug delivery have been assessed for their bio-compatibility via these assays. The cytotoxicity of the thermoresponsive polymers poly(N-isopropylacrylamide), poly(N-vinylcaprolactam) and amphiphilically modified poly(N-vinylcaprolactam) have been analyzed by using two different colorimetric cytotoxicity methods, MTT test and LDH test [58]. Both MTT and LDH tests showed similar trends in the results [58]. Similarly, the biocompatibility assessment of thermoresponsive poly(N-isopropylacrylamide)/polyarginine bioconjugate, developed as nonviral transgene vectors was analyzed through MTT assay [59]. The results indicated a profound impact of neutral polyacrylamide chains in nullifying the cytotoxic effects of charged polyarginine chains, thus rendering the conjugate biocompatible. The cytotoxicity of novel hydrogels prepared via incorporating pH sensitive poly(N-(2-(dimethylamino) ethyl)-methacrylamide) (PDMAEMA) chains to thermoresponsive PNIPAM network was assessed on the HeLa cell lines via MTT assay [60]. The results demonstrated an insignificant decrease in cell viability with the increasing concentration of the polymeric sample from 20 mg/mL to 140 mg/mL [60].

The MTT assay is a sensitive and quantitative colorimetric assay for measuring the activity of enzymes that can reduce MTT to formazan, giving a purple color [61]. This mostly happens in mitochondria by the enzyme succinate dehydrogenase and so the assays are therefore largely a measure of mitochondrial activity. The yellow MTT (3-(4,5-Dimethylthiazol-2-yl)-2,5-diphenyltetrazolium bromide, a tetrazole) is reduced to purple formazan in the mitochondria of living cells [62–64].

A solubilization solution (usually dimethyl sulfoxide) is added to dissolve the insoluble purple formazan product into a colored solution. The absorbance of this colored solution can be quantified by measuring at a certain wavelength (usually 570 nm) by a spectrophotometer. The absorption maximum is dependent on the solvent employed [64].

HeLa cells are a human epithelial cervical cancer and the first human cells, from which a permanent cell line was established. The line was derived from cervical cancer cells taken from Henrietta Lacks, a patient who eventually died of her cancer in 1951 [65]. The cell line was found to be remarkably durable and prolific as illustrated by its contamination of many other cell lines used in research [66]. HeLa cells are termed "immortal" as they can divide an unlimited number of times in a laboratory cell culture plate as long as fundamental cell survival conditions are met [65, 67].

The cytotoxicity assessment of the synthesized thermoresponsive N-isopropylacrylamide-N-vinylpyrrolidone (NIPAM-VP) microgel particles and nanogel particles [68, 69] was carried out on the cultured human cell lines employing the MTT cell viability test.

4.1 Materials

Prepare all the solutions using distilled water and analytical grade reagents. Prepare and store all the reagents at room temperature (unless otherwise indicated). Diligently follow all waste disposal regulations when disposing waste materials.

1. Dulbecco's Modified Eagle's Medium (low glucose) is obtained from Sigma-Aldrich and stored in dark at 4 °C.

2. Trypsin solution is obtained from Gibco (Invitrogen).

3. 10 % fetal calf serum is procured from Pan Biotech GmbH and stored at 4 °C.

4. Penicillin/streptomycin is obtained from Sigma-Aldrich and stored at 4 °C

5. 96-well plate is supplied from Nunc, Thermo scientific.

6. Methylthiazolyldiphenyl-tetrazolium bromide (MTT) [(3-(4,5-Dimethylthiazol-2-yl)-2,5-Diphenyltetrazolium Bromide)] is procured from Sigma-Aldrich. The stock working solution contains 0.5 mg/mL of MTT and stored at 4 °C.

7. Sodium dodecyl sulfate is supplied by Sigma-Aldrich. The stock solution 0.5 % w/v is freshly prepared.

8. Hydrochloric acid is procured from Merck and 25 mM stock solution is used.

9. Isopropyl alcohol is procured from Merck and 90 % solution is used.

10. Dimethylsulfoxide (DMSO) is supplied by Sigma-Aldrich.

4.1.1 Maintenance of Cell Line in the Flask

Cells are split or reefed every 3–4 days (see **Notes 9–12**). In order to obtain the same a typical procedure as mentioned below is followed:

1. The 25 mL flasks are removed from the incubator (maintained at 37 °C) and examined under confocal microscope.

2. The flasks and medium are then placed under laminar hood for some time.

3. In the next step, previous medium is discarded from the flasks and subsequently 400 μL of trypsin solution is added to it.

4. The flasks are kept at 37 °C for 1–2 min after which 600 μL of fresh medium is added to it.

5. The complete cell suspension is then transferred to a falcon tube and centrifuged at 600 rpm for 5 min.

6. Following this, the supernatant is discarded and the pellet obtained is resuspended in 1 mL of the complete medium.

7. 10 mL of the complete medium is added to a new sterile 25 mL volumetric flask and the cell suspension (prepared in **step 6**) is transferred to it.

8. The flasks are recapped and gently shaken to evenly disperse the cells.

9. The new flasks are labelled with the cell line name, the passage, the slit ratio or seeding density, and the date.

10. The flasks are again stored at 37 °C in an incubator till the next reefing of cells or preparation of cell culture plate.

4.1.2 Preparation and Maintenance of Cells in 96-Well Plates

1. The 96-well tissue culture plate and the complete medium are placed under the laminar hood for 5–10 min before starting.

2. 100 µL of cell suspension (stored in the flask at 37 °C *see* Subheading 4.1.1, **item 10**) in an incubator is added to each well using the pipette held at 45° angle about ½ ways down to the well, starting at the top left corner of the plate and working back and forth across the wells to the lower right corner.

3. 150–175 µL of fresh medium is then added into each well using a multichannel pipettor, working from the end of the plate furthest from hand, to minimize the number of times the hand passes over the plate.

4. The plates are then gently shaken to evenly disperse the cells.

5. The plates are finally observed under confocal microscope and then kept in the incubator for 24 h for the next use.

4.2 Methods

Carry out all the procedures at room temperature unless otherwise specified.

4.2.1 Preparation of Different Concentrations of the Synthesized Thermoresponsive and Series of pH Responsive Copolymeric Samples

1. The maximum concentration of 200 µg/mL are prepared by dispersing 1 mg (approximately) of dried polymeric sample in 5 mL of complete medium.

2. The sample prepared (referred to as stock) is sonicated for 10 min to obtain uniform dispersion.

3. In the next step, 500 µL of complete medium is added to six other different falcon tubes (labelled accordingly as 2–7 in Fig. 12).

4. 1 mL of the stock solution is taken in the tube labelled as 1 in Fig. 12 and further dilutions are made by serially transferring 500 µL of higher concentration (200 µg/mL in tube 1) solution to the other tubes already containing 500 µL of complete medium with thorough mixing and sonication after every dilution.

5. The polymeric sample dilutions are hence prepared at seven different concentrations, viz., 200 µg/mL, 100 µg/mL, 50 µg/mL, 25 µg/mL, 12.5 µg/mL, 6.25 µg/mL, and 3.125 µg/mL.

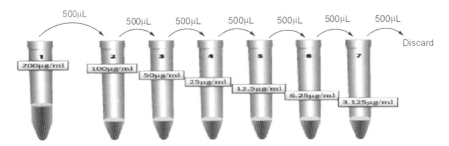

Fig. 12 Preparation of different concentrations of synthesized polymeric samples via serial

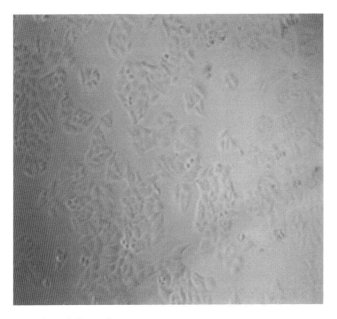

Fig. 13 HeLa cell lines showing 50 % confluence

4.2.2 Administration of the Polymeric Samples at Seven Different Concentrations to the Cultured HeLa Cells in 96-Well Cell Culture Plates

1. The 96-well cell culture plate is found ready for inoculation after 24 h of its incubation at 37 °C as when observed under confocal microscope it showed 50 % confluence (*see* Fig. 13).

2. Following this, 300 μL of various concentrations of the thermoresponsive polymeric samples hence made are added in triplicates to each well and the plates are labelled accordingly, as shown in Fig. 14.

3. The untreated cells are used as a negative control (100 % cell viability) and a mitochondrial activity inhibiting drug is used as positive control (0 % cell viability).

4. Later, the plates are incubated at 37 °C for 24 h to facilitate the cellular uptake of test material which is inoculated along with the nutrient media.

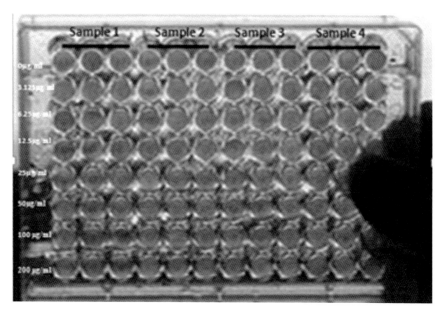

Fig. 14 HeLa cell lines incubated with nanoparticles of defined concentrations

4.2.3 MTT Assay for Cytotoxicity Analysis

1. Following treatment with various concentrations of polymeric samples, the incubated 96-well cell culture plates are examined under confocal microscope to observe the morphological changes in the HeLa cell lines (if any).

2. The culture medium is then replaced with serum-free medium containing 10 µL MTT solution (0.5 mg/mL) in each well. The solution is mixed by tapping gently on the side of the tray or by shaking briefly on an orbital shaker.

3. Cultures are incubated for an additional 3 h at 37 °C.

4. The blue MTT formazan is dissolved in the specified MTT buffer (0.5 % SDS w/v, 25 mM HCl in 90 % isopropyl alcohol).

5. Measure the absorbance values using UV spectrophotometer with a test wavelength of 570 nm and a reference wavelength of 630 nm to obtain sample signal (OD570–OD630).

6. The untreated cells are used as a negative control (i.e., 100 % viable) and a mitochondrial activity inhibiting drug as a positive control (100 % cell death). All values from the experiment are correlated with this set of data.

7. A graphical representation of percentage cell viability and concentration of thermoresponsive polymer is made to assess the biocompatibility of the particles (*see* Figs. 15 and 16). The basic layout of the MTT assay performed for testing the cytotoxicity of the thermosensitive polymers is shown in Fig. 17.

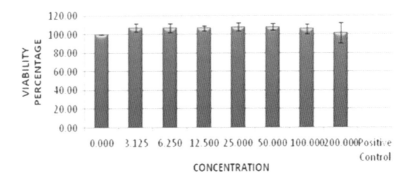

Fig. 15 Viability percentages of HeLa cells at different concentrations of NIPAM-VP microgel MG1 particles

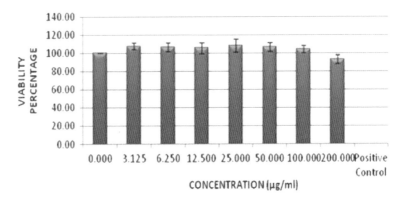

Fig. 16 Viability percentage of HeLa cells at different concentrations of NIPAM-VP Nanogel NG1

5   Notes

1. The freeze drier instrument Ilshin model-TFD5505 is used and operated at condenser temperature −80 °C and vacuum pressure less than 5 mTorr.

2. The QELS measurements of the particles are done by Photocor-FC model-1135P instrument. The light source is an argon ion laser operating at 633 nm and detection of diffracted light is measured at an angle of 90° to an incident laser beam. Diffractogram is processed by Photocor and Dynal software.

3. SEM analyses are performed on JEOL model JSM 840 instrument operating at voltage of 10–15 kV.

4. TEM analyses are performed on FEI Philips Morgani model 268D instrument. The operating voltage ranges from 10 kV to 300 kV with the magnification of the image going upto 300,000 times.

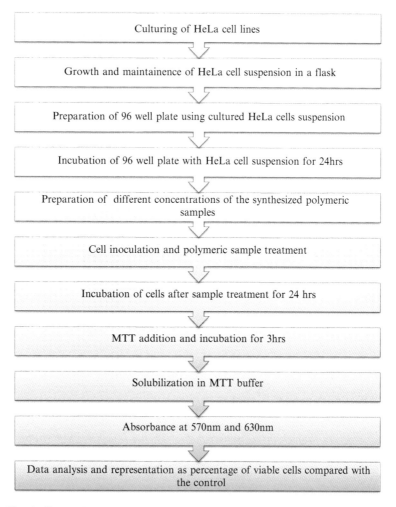

Fig. 17 The basic layout of the MTT assay performed for testing the cytotoxicity of the thermosensitive polymers.

5. The pH 7.4 is ensured by dipping pH electrode of pre-calibrated pH meter in the buffer solution and minor adjustment of pH is done with aqueous solution of 0.2 M NaOH or 0.2 M orthophosphoric acid.

6. The samples for UV analysis are filtered through 0.45 μ glass nylon filers.

7. UV analysis is done using a double beam UV spectrophotometer instrument Systronic Model-AU 2700 (India).

8. Place plain phosphate buffer (at a pH of 7.4) in both front and rear cavities of UV–Vis spectrophotometer. The instrument is instructed to set the difference in absorbance equal to zero. Keep the blank (plain phosphate buffer pH of 7.4) in the rear cavity and place the samples in front cavity and record the absorbance.

9. The UV light in the hood is turned on and allowed to run for at least 10 min before starting.

10. The complete media is pre-warmed in a water bath maintained at 37 °C.

11. All surfaces and hands were sterilized with 70 % alcohol before initiating any experiment.

12. Culturing and maintenance of mammalian cell lines HeLa involves the formation of a suspension of HeLa cell lines in a flask, maintained at 37 °C under 7 % CO_2 in low-glucose DMEM supplemented with 10 % fetal calf serum and penicillin/streptomycin, which are reefed every third to fourth day so as to gauge the cell growth [70].

Acknowledgement

This work was supported by Department of Biotechnology (DBT) New Delhi, India grant no. BT/PR8918/NNT/28/05/2007. The authors thank Swati Aerry and Mitasha Bharadwaj for this work and Dr. Y. Singh (Scientist "G", IGIB, Delhi) for carrying out the cytotoxicity work in his lab.

References

1. Roy D, Cambre JN, Sumerlin BS (2010) Future perspectives and recent advances in stimuli-responsive materials. Prog Polym Sci 35:278–301

2. Mano JF (2008) Stimuli-responsive polymeric systems for biomedical applications. Adv Eng Mater 10:515–527

3. Kamaly N, Xiao Z, Valencia PM, Radovic-Moreno AF, Farokhzad OC (2012) Targeted polymeric therapeutic nanoparticles: design, development and clinical translation. Chem Soc Rev 41:2971–3010

4. Bae YH, Okano T, Hsu R, Kim SW (1987) Thermo-sensitive polymers as on-off switches for drug release. Makromol Chem Rapid Commun 8:481–485

5. Jeong B, Bae YH, Lee DS, Kim SW (1997) Biodegradable block copolymers as injectable drug-delivery systems. Nature 388:860–862

6. Pillai O, Panchagnula R (2001) Polymers in drug delivery. Curr Opin Chem Biol 5:447–451

7. Aguilar M, Elvira C, Gallardo A, Vázquez B, Román J (2007) Smart polymers and their applications as biomaterials. In: Ashammakhi N, Reis R, Chiellini E (ed) Topics in tissue engineering, Vol 3, University of Oulu (Expert issues e-books), Finland, pp. 1–27

8. Li S (2010) Smart polymer materials for biomedical applications. Nova Science Publishers, Incorporated

9. Gerasimov OV, Boomer JA, Qualls MM, Thompson DH (1999) Cytosolic drug delivery using pH- and light-sensitive liposomes. Adv Drug Deliv Rev 38:317–338

10. Alvarez-Lorenzo C, Bromberg L, Concheiro A (2009) Light-sensitive intelligent drug delivery systems. Photochem Photobiol 85:848–860

11. Chung JE, Yokoyama M, Yamato M, Aoyagi T, Sakurai Y, Okano T (1999) Thermo-responsive drug delivery from polymeric micelles constructed using block copolymers of poly(N-isopropylacrylamide) and poly (butylmethacrylate). J Control Release 62: 115–127

12. Li Y, Pan S, Zhang W, Du Z (2009) Novel thermo-sensitive core-shell nanoparticles for targeted paclitaxel delivery. Nanotechnology 20:065104

13. Chung JE, Yokoyama M, Okano T (2000) Inner core segment design for drug delivery

control of thermo-responsive polymeric micelles. J Control Release 65:93–103

14. Satturwar P, Eddine MN, Ravenelle F, Leroux J-C (2007) pH-responsive polymeric micelles of poly (ethylene glycol)-b-poly (alkyl (meth) acrylate-co-methacrylic acid): influence of the copolymer composition on self-assembling properties and release of candesartan cilexetil. Eur J Pharm Biopharm 65:379–387

15. Na K, Lee KH, Bae YH (2004) pH-sensitivity and pH-dependent interior structural change of self-assembled hydrogel nanoparticles of pullulan acetate/oligo-sulfonamide conjugate. J Control Release 97:513–525

16. Chen S-C, Wu Y-C, Mi F-L, Lin Y-H, Yu L-C, Sung H-W (2004) A novel pH-sensitive hydro-gel composed of N, O-carboxymethyl chitosan and alginate cross-linked by genipin for protein drug delivery. J Control Release 96:285–300

17. Hrubý M, Koňák Č, Ulbrich K (2005) Polymeric micellar pH-sensitive drug delivery system for doxorubicin. J Control Release 103:137–148

18. Sawahata K, Hara M, Yasunaga H, Osada Y (1990) Electrically controlled drug delivery system using polyelectrolyte gels. J Control Release 14:253–262

19. Kwon IC, Bae YH, Okano T, Kim SW (1991) Drug release from electric current sensitive polymers. J Control Release 17:149–156

20. Yuk SH, Cho SH, Lee HB (1992) Electric current-sensitive drug delivery systems using sodium alginate/polyacrylic acid composites. Pharm Res 9:955–957

21. Kost J, Leong K, Langer R (1988) Ultrasonically controlled polymeric drug deliv-ery. Paper presented at Makromolekulare Chemie. Macromolecular Symposia, 1988

22. Kost J, Leong K, Langer R (1989) Ultrasound-enhanced polymer degradation and release of incorporated substances. Proc Natl Acad Sci U S A 86:7663–7666

23. Ito Y, Casolaro M, Kono K, Imanishi Y (1989) An insulin-releasing system that is responsive to glucose. J Control Release 10:195–203

24. Shiino D, Murata Y, Kataoka K et al (1994) Preparation and characterization of a glucose-responsive insulin-releasing polymer device. Biomaterials 15:121–128

25. Hisamitsu I, Kataoka K, Okano T, Sakurai Y (1997) Glucose-responsive gel from phenylbo-rate polymer and poly (vinyl alcohol): prompt response at physiological pH through the interaction of borate with amino group in the gel. Pharm Res 14:289–293

26. Dong-June C, Yoshihiro I, Yukio I (1992) An insulin-releasing membrane system on the basis of oxidation reaction of glucose. J Control Release 18:45–53

27. Ulijn RV (2006) Enzyme-responsive materials: a new class of smart biomaterials. J Mater Chem 16:2217–2225

28. Thornton PD, McConnell G, Ulijn RV (2005) Enzyme responsive polymer hydrogel beads. Chem Commun 47:5913–5915

29. Toledano S, Williams RJ, Jayawarna V, Ulijn RV (2006) Enzyme-triggered self-assembly of peptide hydrogels via reversed hydrolysis. J Am Chem Soc 128:1070–1071

30. Miyata T, Asami N, Uragami T (1999) Preparation of an antigen-sensitive hydrogel using antigen-antibody bindings. Macromolecules 32:2082–2084

31. Lu ZR, Kopečková P, Kopeček J (2003) Antigen responsive hydrogels based on polym-erizable antibody Fab′ fragment. Macromol Biosci 3:296–300

32. Zhang R, Bowyer A, Eisenthal R, Hubble J (2007) A smart membrane based on an antigen-responsive hydrogel. Biotechnol Bioeng 97:976–984

33. Koo AN, Lee HJ, Kim SE et al (2008) Disulfide-cross-linked PEG-poly (amino acid)s copolymer micelles for glutathione-mediated intracellular drug delivery. Chem Commun:6570–6572

34. Tsarevsky NV, Matyjaszewski K (2005) Combining atom transfer radical polymerization and disulfide/thiol redox chemistry: a route to well-defined (bio) degradable polymeric materi-als. Macromolecules 38:3087–3092

35. He C, Kim SW, Lee DS (2008) In situ gelling stimuli-sensitive block copolymer hydrogels for drug delivery. J Control Release 127:189–207

36. Gil ES, Hudson SM (2004) Stimuli-responsive polymers and their bioconjugates. Prog Polym Sci 29:1173–1222

37. Dirk S (2006) Thermo- and pH-responsive polymers in drug delivery. Adv Drug Deliv Rev 58:1655–1670

38. Klouda L, Mikos AG (2008) Thermoresponsive hydrogels in biomedical applications. Eur J Pharm Biopharm 68:34–45

39. Schild HG (1992) Poly(N-isopropylacrylamide): experiment, theory and application. Prog Polym Sci 17:163–249

40. Yan H, Tsujii K (2005) Potential application of poly(N-isopropylacrylamide) gel containing polymeric micelles to drug delivery systems. Colloids Surf B Biointerfaces 46:142–146

41. Fitzpatrick SD, Jafar Mazumder MA, Muirhead B, Sheardown H (2012) Development of injectable, resorbable drug-releasing copolymer scaffolds for minimally invasive sustained ophthalmic therapeutics. Acta Biomater 8:2517–2528

42. Karir T, Sarma HD, Samuel G, Hassan PA, Padmanabhan D, Venkatesh M (2013) Preparation and evaluation of radioiodinated thermoresponsive polymer based on poly(N-isopropyl acrylamide) for radiotherapy. J Appl Polym Sci 130:860–868

43. Chen Y-Y, Wu H-C, Sun J-S, Dong G-C, Wang T-W (2013) Injectable and thermoresponsive self-assembled nanocomposite hydrogel for long-term anticancer drug delivery. Langmuir 29:3721–3729

44. Jiang B, Larson JC, Drapala PW, Pérez-Luna VH, Kang-Mieler JJ, Brey EM (2012) Investigation of lysine acrylate containing poly (N-isopropylacrylamide) hydrogels as wound dressings in normal and infected wounds. J Biomed Mater Res B Appl Biomater 100:668–676

45. Tan H, Ramirez CM, Miljkovic N, Li H, Rubin JP, Marra KG (2009) Thermosensitive injectable hyaluronic acid hydrogel for adipose tissue engineering. Biomaterials 30:6844–6853

46. Luo L, Ranger M, Lessard DG et al (2004) Novel amphiphilic diblock copolymer of low molecular weight poly(N-vinylpyrrolidone)-block-poly(d, l-lactide): synthesis, characterization and micellization. Macromolecules 37:4008–4013

47. Haaf F, Sanner A, Straub F (1985) Polymers of N-vinylpyrrolidone: synthesis, characterization and uses. Polymer J 17:143–152

48. D'Souza AJM, Schowen RL, Topp EM (2004) Polyvinylpyrrolidone–drug conjugate: synthesis and release mechanism. J Control Release 94:91–100

49. Zhang L, Liang Y, Meng L, Lu X, Liu Y (2007) Preparation and PCR-amplification properties of a novel amphiphilic poly(N-vinylpyrrolidone) (PVP) copolymer. Chem Biodivers 4:163–174

50. Indian Pharmacopoeia Delhi, Government of India, Ministry of Health and Family Welfare: Published by the controller of Publication; 1996

51. Niles AL, Moravec RA, Riss TL (2009) In vitro viability and cytotoxicity testing and same-well multi-parametric combinations for high throughput screening. Curr Chem Genomics 3:33–41

52. Cook JA, Mitchell JB (1989) Viability measurements in mammalian cell systems. Anal Biochem 179:1–7

53. Weyermann J, Lochmann D, Zimmer A (2005) A practical note on the use of cytotoxicity assays. Int J Pharm 288:369–376

54. Fotakis G, Timbrell JA (2006) In vitro cytotoxicity assays: comparison of LDH, neutral red, MTT and protein assay in hepatoma cell lines following exposure to cadmium chloride. Toxicol Lett 160:171–177

55. Korzeniewski C, Callewaert DM (1983) An enzyme-release assay for natural cytotoxicity. J Immunol Methods 64:313–320

56. Decker T, Lohmann-Matthes M-L (1988) A quick and simple method for the quantitation of lactate dehydrogenase release in measurements of cellular cytotoxicity and tumor necrosis factor (TNF) activity. J Immunol Methods 115:61–69

57. Jurišić V, Bumbaširević V (2008) In vitro assays for cell death determination. Arch Oncol 16:49–54

58. Vihola H, Laukkanen A, Valtola L, Tenhu H, Hirvonen J (2005) Cytotoxicity of thermosensitive polymers poly(N-isopropylacrylamide), poly(N-vinylcaprolactam) and amphiphilically modified poly(N-vinylcaprolactam). Biomaterials 26:3055–3064

59. Cheng N, Liu W, Cao Z et al (2006) A study of thermoresponsive poly(N-isopropylacrylamide)/polyarginine bioconjugate non-viral transgene vectors. Biomaterials 27:4984–4992

60. Wang Z-C, Xu X-D, Chen C-S et al (2008) Study on novel hydrogels based on thermosensitive PNIPAAm with pH sensitive PDMAEMA grafts. Colloids Surf B Biointerfaces 67:245–252

61. Barltrop JA, Owen TC, Cory AH, Cory JG (1991) 5-(3-carboxymethoxyphenyl)-2-(4,5-dimethylthiazolyl)-3-(4-sulfophenyl)tetrazolium, inner salt (MTS) and related analogs of 3-(4,5-dimethylthiazolyl)-2,5-diphenyltetrazolium bromide (MTT) reducing to purple water-soluble formazans as cell-viability indicators. Bioorg Med Chem Lett 1:611–614

62. Mosmann T (1983) Rapid colorimetric assay for cellular growth and survival: application to proliferation and cytotoxicity assays. J Immunol Methods 65:55–63

63. Hansen MB, Nielsen SE, Berg K (1989) Re-examination and further development of a precise and rapid dye method for measuring cell growth/cell kill. J Immunol Methods 119:203–210

64. Riss TL, Moravec RA, Niles AL, Benink HA, Worzella TJ, Minor L (2013) Cell viability assays. In: Sittampalam GS, Gal-Edd N, Arkin M et al (eds) Assay guidance manual. Bethesda, MD: Eli Lilly & Company and the National

Center for Advancing Translational Sciences; Available from: http://www.ncbi.nlm.nih.gov/books/NBK144065/

65. Masters JR (2002) HeLa cells 50 years on: the good, the bad and the ugly. Nat Rev Cancer 2:315–319

66. Capes-Davis A, Theodosopoulos G, Atkin I et al (2010) Check your cultures! A list of cross-contaminated or misidentified cell lines. Int J Cancer 127:1–8

67. Lucey BP, Nelson-Rees WA, Hutchins GM (2009) Henrietta lacks, HeLa cells, and cell culture contamination. Arch Pathol Lab Med 133:1463–1467

68. Aerry S, De A, Kumar A, Saxena A, Majumdar D, Mozumdar S (2013) Synthesis and characterization of thermoresponsive copolymers for drug delivery. J Biomed Mater Res A 101(7):2015–26

69. Aerry S (2010) Synthesis and characterization of polymers and polymeric nanoparticles for applications in drug delivery. Ph.D. thesis submitted at the Department of Chemistry, University of Delhi, 2010

70. Rahbari R, Sheahan T, Modes V, Collier P, Macfarlane C, Badge RM (2009) A novel L1 retrotransposon marker for HeLa cell line identification. Biotechniques 46:277

Chapter 5

Recombinant Stem Cells as Carriers for Cancer Gene Therapy

Yu-Lan Hu and Jian-Qing Gao

Abstract

Recent studies have shown the ability of mesenchymal stem cells (MSCs) to migrate toward and engraft into the tumor sites, which provides a potential for their use as carriers for cancer gene therapy. Here, we describe the strategies of using MSCs as carriers for cancer gene therapy using a nonviral transfection method.

Key words Carrier, Mesenchymal stem cells, Nonviral transfection, Cancer

1 Introduction

Targeted delivery of anticancer agents is one of promising fields in anticancer therapy. A major drawback of anticancer agents is their lack of selectivity for tumor tissue, which causes severe side effects and results in low therapeutic efficiency. Therefore, tumor-targeting approaches have been developed for improved efficacy and for minimizing systematic toxicity by altering biodistribution profiles of anticancer agents.

Cell-based therapies are emerging as promising therapeutic options for cancer treatment. However, the clinical application of differentiated cells is hindered by the difficulty in obtaining a large number of cells, their lack of ability to expand in vitro, as well as poor engraftment efficacy to targeted tumor sites. Mesenchymal stem cells (MSCs) have been attractive cell therapy vehicles for the delivery of agents into tumor cells because of their capability of self-renewal, relative ease of isolation and expansion in vitro, and homing capacity allowing them to migrate toward and engraft into the tumor [1]. Several studies have provided evidence supporting the rationale for genetically modified MSCs to deliver therapeutic cytokines directly into the tumor microenvironment to produce high concentrations of antitumor proteins at the tumor sites, which

Kewal K. Jain (ed.), *Drug Delivery System*, Methods in Molecular Biology, vol. 1141,
DOI 10.1007/978-1-4939-0363-4_5, © Springer Science+Business Media New York 2014

have been shown to inhibit tumor growth in experimental animal models. The antitumor effects of intravenous injections of gene-modified MSCs have been demonstrated in lung, brain, and subcutaneous tumors [2–5]. Widely used nonviral delivery approaches are calcium phosphate, cationized liposomes, noisomes, and cationic polymers [6–9]. Transduction of MSCs by polyethylenimine with the molecular weight of 25 kDa (PEI25kDa) is highly efficient [10]. Nevertheless, PEI25kDa is associated with high cytotoxicity. Thus, the nonviral vector, spermine-pullulan, was used for the transfection of MSCs. This chapter describes the experimental methods for cancer gene therapy using MSCs, the culture of MSCs from murine bone marrow, and the nonviral transfection of MSCs.

2   Materials

2.1   Reagents

Dulbecco's Modified Eagle's Medium (DMEM) (Gibco BRL, Gaithersberg, MD, USA).

Fetal bovine serum (FBS) (Gibco BRL, Gaithersberg, MD, USA).

Penicillin and streptomycin (Gibco BRL, Gaithersberg, MD, USA).

Trypsin (Gibco BRL, Gaithersberg, MD, USA).

[3-(4, 5-dimethylthiazol-2-yl)-2,5-diphenyltetrazolium bromide] (MTT) (Sigma-Aldrich Chemical Co., St. Louis, MO).

Plasmids (pGL3) (Promega, Madison, WI, USA).

Plasmid (pEGFP-N1) (Clontech, Palo Alto, CA).

Plasmid (IL-12) was a generous gift from Dr. Shinsaku Nakagawa (Osaka University, Japan).

Luciferase Assay Kit (Beyontime Co., China).

BCA Protein Assay Kit (Beyontime Co., China).

Spermine-pullulan (SP) was synthesized by our laboratory using the method described previously and confirmed by the conventional elemental analysis [11] (see **Note 1**).

Anti-CD45 and anti-CD90 antibody (BioLegend, San Diego, CA, USA).

Anti-CD34 antibody (Santa Cruz Biotechnology, Inc, Santa Cruz, CA, USA).

Anti-CD73 antibody (BD Pharmingen, Heidelberg, Germany).

FITC-labeled secondary antibodies (Boster, Wuhan, Hubei, China).

2.2   Animals

SD (Sprague–Dawley) male rats (3 weeks old, Zhejiang University Experimental Animal Center, China).

Female C57BL/6 mice (6–8 weeks old, Zhejiang University Experimental Animal Center, China).

3 Methods

3.1 Isolation and Culture of MSCs

1. All animal experiments should comply with national laws and institutional regulations.

2. Euthanize an SD rat by cervical dislocation.

3. Remove the bones from the SD rats.

4. Dissected cleaning of attached muscles from the bone.

5. A needle attached to a 1-ml syringe containing complete DMEM is inserted into one end of each bone, and the medium is forced through the bone cavity to expel the marrow from the other end of the bone.

6. The medium containing bone marrow is passed through a 70-μm sterile filter.

7. The cell suspension is placed into a 10-cm dish and cultured at 37 °C in humidified atmosphere containing 95 % air and 5 % CO_2.

8. When the cells of the first passage became sub-confluent, usually 7–10 days after seeding, the cells are split 1:3 onto new 10-cm dishes.

9. This process is repeated until four to six passages are completed. The MSCs of passage four are examined for the morphology and expression of CD73, CD90, CD34, and CD45 by flow cytometry (Fig. 1). These cells at sub-confluence are used for all experiments.

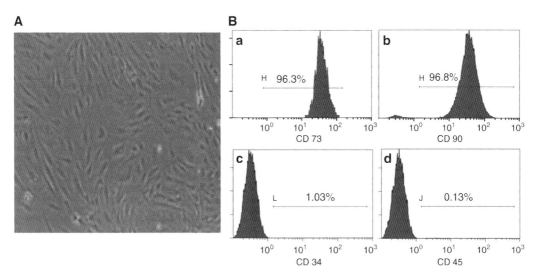

Fig. 1 Characterization of MSCs. (**a**) Morphology observation of MSCs of passage four. (**b**) Expression of CD73, CD90, CD34, and CD45 of MSC of passage 4 by flow cytometry

3.2 Nonviral Gene Transfection of MSCs

1. Prior to transduction, MSCs were seeded in a 24-well plate at a density of 1×105 cells/well, and the cells are incubated for 24 h at 37 °C with 5 % CO_2.

2. Vector/PGL-3 complexes were prepared by adding cationic polymer solution to equal volumes of DNA solution with gentle vortex mixing and incubated at room temperature for 20 min (*see* **Note 2** and **Note 3**).

3. The original cell culture media was replaced with the complex solution and an additional of 500 μl Opti-MEM in each well.

4. After incubation for 4–6 h at 37 °C, the transfection medium was replaced with fresh growth medium containing 15 % FBS and then the cells were incubated for 18–24 h.

The luciferase assay was carried out according to the manufacture's instruction (Promega, USA) (*see* **Note 4**).

3.3 Cytotoxicity Evaluation of Cationic Polymer

1. The spermine solutions were filtering through 0.22-μm aseptic filter membranes before they were incubated with MSCs.

2. For cell viability assay, MSCs were seeded into 96-well plates at a density of 5×10^3–1×10^4 cells/well, and the cells are incubated for 24 h at 37 °C with 5 % CO_2.

3. After 24 h, different concentrations of CP and PEI were added with 200-μl fresh serum-free DMEM to replace the culture medium.

4. MSCs were incubated at 37 °C in a humidified 5 % CO_2-containing atmosphere for 4–6 h. Then the medium was replaced by fresh DMEM with 10 % fetal bovine serum.

5. After being incubated for 24 h, 20 μl of MTT solution (5 mg/ml) was added to each well and the absorbance was measured according to the manufacture's instruction. The cell viability was calculated as following:

$$Cell\ viability\ (\%) = \frac{OD570\ (sample)}{OD570\ (control)} \times 100$$

where OD570 (sample) represents a measurement from the group treated with vectors and OD570 (control) represents the group treated with PBS buffer only.

3.4 Gene-Engineered MSCs for Cancer Therapy

1. B16F10 cells (1×10^5 cells in 0.2-ml PBS per mouse) were injected into C57BL/6 mice via the tail vein.

2. Following the B16F10 cells injection, mice were randomly divided into different groups.

3. MSCs are transfected with IL-12 using the method described in Subheading 3.2 (*see* **Note 5**).

4. 10 days after the establishment of lung metastases in C57BL/6 mice, intravenous inoculation of 0.2-ml IL-12-engineered MSCs (5×10^6/ml), MSCs-GFP (5×10^6/ml), vector/IL-12 (125 μg/ml IL-12 DNA), or PBS (6 animals/group) was performed via the lateral tail vein.

5. All mice were sacrificed at 21 days after tumor cell injection.

6. Lungs were removed from the mice and the metastatic foci, which size ranges between 0.1 and 0.5 mm in diameter, at lung surfaces were photographed and counted and the lung weight was measured.

7. The lung from MSCs-IL-12- and MSCs-GFP-treated group was taken out and undergone paraffin section.

8. Terminal deoxyribonucleotidyl transferase-mediated dUTP nick end labeling (TUNEL) staining was performed using a TUNEL Assay Kit (Roche) according to the manufacturer's instructions.

4   Notes

1. The cationic polymer spermine-pullulan was synthesized in our lab [11]. Other vectors, such as chitosan linked with PEI and β-cyclodextrin linked with polyethylenimine, could also be used.

2. The quality of the DNA can affect the transfection efficiency. Impurities such as high EDTA concentration, RNA contamination, and presence of endotoxins can interfere with complex formation between the polymer and DNA.

3. The N/P ratio, defined as the ratio of moles of the amine groups of cationic polymer to moles of phosphates of DNA, will affect the transfection efficiency of cationic polymer. Therefore, the different N/P ratio should be used to examine the transfection efficiency. For the transfection of MSCs, the optimal N/P ratio of spermine-pullulan is 4.

4. The transfection efficiency could also be examined by percentage of EGFP expression quantified using flow cytometry.

5. The plasmid IL-12 could be replaced by other tumor therapeutic plasmid such as tumor necrosis factor (TNF)-related apoptosis-inducing ligand (TRAIL) [12] or herpes simplex virus thymidine kinase (HSV-TK) combined with ganciclovir (GCV) [13].

Acknowledgement

This work was financially supported by National Natural Science Foundation of China (30873173 and 81273441 to Jian-Qing Gao, 81001410 to Yu-Lan Hu), Zhejiang Provincial Natural

Science Foundation of China (Y13H300002 to Yu-Lan Hu), and Zhejiang Provincial Program for the Cultivation of High-Level Innovative Health Talents.

References

1. Dennis JE, Cohen N, Goldberg VM et al (2004) Targeted delivery of progenitor cells for cartilage repair. J Orthop Res 22: 735–741

2. Studeny M, Marini FC, Dembinski JL et al (2004) Mesenchymal stem cells: potential precursors for tumor stroma and targeted-delivery vehicles for anticancer agents. J Natl Cancer Inst 96:1593–1603

3. Nakamura K, Ito Y, Kawano Y et al (2004) Antitumor effect of genetically engineered mesenchymal stem cells in a rat glioma model. Gene Ther 11:1155–1164

4. Nakamizo A, Marini F, Amano T et al (2005) Human bone marrow derived mesenchymal stem cells in the treatment of gliomas. Cancer Res 65:3307–3318

5. Kucerova L, Altanerova V, Matuskova M et al (2007) Adipose tissue-derived human mesenchymal stem cells mediated prodrug cancer gene therapy. Cancer Res 67:6304–6313

6. Yang X, Walboomers XF, van den Dolder J et al (2008) Non-viral bone morphogenetic protein 2 transfection of rat dental pulp stem cells using calcium phosphate nanoparticles as carriers. Tissue Eng Part A 14:71–81

7. Bisht S, Bhakta G, Mitra S et al (2005) pDNA loaded calcium phosphate nanoparticles: highly efficient non-viral vector for gene delivery. Int J Pharm 288:157–168

8. Ding W, Izumisawa T, Hattori Y et al (2009) Non-ionic surfactant modified cationic liposomes mediated gene transfection in vitro and in the mouse lung. Biol Pharm Bull 32: 311–315

9. Huang YZ, Gao JQ, Chen JL et al (2006) Cationic liposomes modified with non-ionic surfactants as effective non-viral carrier for gene transfer. Colloids Surf B Biointerfaces 49: 158–164

10. Pack DW, Hoffman AS, Pun S et al (2005) Design and development of polymers for gene delivery. Design and development of polymers for gene delivery. Nat Rev Drug Discov 4: 581–593

11. Jo J, Ikai T, Okazaki A et al (2007) Expression profile of plasmid DNA by spermine derivatives of pullulan with different extents of spermine introduced. J Control Release 118:389–398

12. Hu YL, Huang B, Zhang TY et al (2012) Mesenchymal stem cells as a novel carrier for targeted delivery of gene in cancer therapy based on nonviral transfection. Mol Pharm 9: 2698–2709

13. Zhang TY, Huang B, Yuan ZY et al (2013) Gene recombinant bone marrow mesenchymal stem cells as a tumor-targeted suicide gene delivery vehicle in pulmonary metastasis therapy using non-viral transfection. Nanomedicine 2013 10:257–67

Chapter 6

Microfluidic-Based Manufacture of siRNA-Lipid Nanoparticles for Therapeutic Applications

Colin Walsh, Kevin Ou, Nathan M. Belliveau, Tim J. Leaver, Andre W. Wild, Jens Huft, Paulo J. Lin, Sam Chen, Alex K. Leung, Justin B. Lee, Carl L. Hansen, Robert J. Taylor, Euan C. Ramsay, and Pieter R. Cullis

Abstract

A simple, efficient, and scalable manufacturing technique is required for developing siRNA-lipid nanoparticles (siRNA-LNP) for therapeutic applications. In this chapter we describe a novel microfluidic-based manufacturing process for the rapid manufacture of siRNA-LNP, together with protocols for characterizing the size, polydispersity, RNA encapsulation efficiency, RNA concentration, and total lipid concentration of the resultant nanoparticles.

Key words Lipid nanoparticle, Microfluidics, siRNA, Nanoparticle manufacture, Solid-core, LNP, siRNA-LNP, NanoAssemblr, Nanoparticle formulation

---

1 Introduction

There are numerous manufacturing methods for the preparation of lipid nanoparticles (LNP) for therapeutic applications [1–4]. The three main techniques are sonication, extrusion, and microfluidization [3, 4]. Sonication and extrusion differ only in the method used to alter LNP morphology and reduce particle size. A hydrated multilamellar vesicle (MLV) suspension is obtained after rehydrating dried lipid films in aqueous solution. Sonication using a bath sonicator or a probe sonicator reduces particle size and produces large unilamellar vesicles (LUV ~100 nm) or small unilamellar vesicles (SUV < 100 nm). In contrast, extrusion produces LUV or SUV by forcing the MLV suspensions through membranes of defined pore size for repeated cycles. In both cases, the size, morphology (LUV versus SUV), and polydispersity of the resultant LNP are dependent on lipid composition, lipid concentration, temperature, pH, volume and sonication time/power/tuning (sonication), or membrane pore size/number of extrusion cycles

Kewal K. Jain (ed.), *Drug Delivery System*, Methods in Molecular Biology, vol. 1141, DOI 10.1007/978-1-4939-0363-4_6, © Springer Science+Business Media New York 2014

(extrusion) [3, 4]. Sonication and extrusion are sensitive techniques that suffer from batch-to-batch variation and are not easily scaled.

Microfluidization provides an alternative LNP manufacturing technique. This method uses high pressure (up to 10,000 psi) to force a lipid suspension, which has been divided into two streams, through a small orifice into an interaction chamber where the two streams collide to produce LNP. The LNP are further reduced in size by cavitation, shear, and impact [3]. Similar to sonication and extrusion, particle size is dependent on LNP parameters such as lipid components, concentration, etc. In addition, pressure, size of the interaction chamber, and the number of microfluidization cycles influence LNP size. Microfluidization can reproducibly produce large volumes of LNP; however, despite reasonable encapsulation efficiencies (>75 %), the high shear force can damage encapsulated material. Moreover, problems with scale-up, material loss, and contamination limit the applicability of this LNP manufacturing method [3, 4].

Traditional manufacturing processes for the production of LNP are problematic [5]. LNP are formulated via bulk mixing and suffer from microscopic variations in the speed and local concentration of lipid components, resulting in nanoparticle heterogeneity, batch-to-batch variability, and operator-dependent formulation characteristics [5]. Further, these "one-pot" manufacturing processes prevent precise control over temporal and spatial concentration of LNP components. Microfluidics provides a versatile and elegant approach to enable the simple and consistent production of LNP [6]. Microfluidic systems typically operate in the regime of laminar flow and afford precise, reproducible, and controlled mixing at the nanoliter scale, enabling conditions ideal for the formulation of nanoparticles and microparticles [7–10]. Further, microfluidic synthesis offers important practical advantages including low reagent consumption, parallelization, and automation, making it ideally suited to high-throughput optimization and scale-up.

The NanoAssemblr™ is a recently developed microfluidic-based platform for the manufacture of nanoparticles such as liposomes, oil-in-water nanoemulsions, and short interfering RNA (siRNA) lipid nanoparticles (siRNA-LNP) (Precision NanoSystems Inc., Vancouver, Canada). The NanoAssemblr™ platform enables efficient encapsulation of siRNA in LNP, and data generated clearly demonstrates microfluidic production to be superior to current manufacturing methods.

The NanoAssemblr™ platform comprises a benchtop instrument with computer software that controls the flow of fluids through a microfluidic mixing device. The microfluidic device consists of a Y-junction followed by a mixing region where staggered herringbone structures induce rapid mixing by chaotic advection (Fig. 1). The NanoAssemblr™ benchtop instrument uses syringe pumps to introduce a lipid/ethanol solution and an aqueous solution

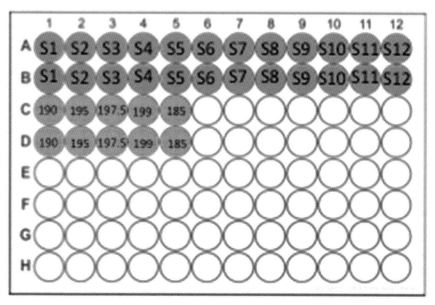

Fig. 1 Schematic overview of the NanoAssemblr™ microfluidic mixer design. (*A*) and (*B*) are inlets where the aqueous siRNA solution and the ethanolic lipid solution are introduced into the microfluidic channels. (*C*) is an enlargement of the staggered herringbone mixer (SHM) used to promote rapid mixing of the aqueous and ethanol solutions. (*D*) is the outlet where formulated siRNA-lipid nanoparticles (siRNA-LNP) are collected

of siRNA into the device (inlet A and inlet B in Fig. 1) at flow rates of 4–12 mL/min. As the two streams pass through the herringbone mixer, the rapid mixing of the ethanol and aqueous streams results in a reduced solubility of lipids (Fig. 1 (C)), causing a state of supersaturation that drives the spontaneous aggregation of lipid into well-defined nanoparticles that simultaneously encapsulate the siRNA via electrostatic interaction with the cationic lipid component of the LNP. Following microfluidic manufacture, the residual ethanol is removed, the buffer exchanged, and the pH adjusted to pH 7.4 by dialysis. The final siRNA-LNP formulation is subsequently characterized according to particle size and polydispersity, RNA encapsulation efficiency, RNA concentration, and lipid concentration.

Microfluidic-based manufacture results in siRNA-LNP formulations with superior characteristics to those produced using traditional methods, such as the preformed vesicle (PFV) or spontaneous vesicle formation (SVF) techniques. Microfluidic manufacture mediates (1) siRNA encapsulation efficiencies of 100 % versus 65–95 % for PFV and SFV methods; (2) defined particle size including the formulation of limit-size LNP ($\geq$20 nm), which are unobtainable with either the PFV or SVF methods; and (3) LNP comprising ionizable cationic lipid in excess of 60 mol% compared to a maximum of 40 mol% achievable with the PFV approach. Collectively, this translates to reduced cost of goods, improved biodistribution, and increased potency in vitro and in vivo.

2   Materials (*See* Note 1)

2.1 Aqueous siRNA Formulation Solution	1. Lyophilized siRNA.
	2. RNA hydration buffer: 10 mM sodium acetate, pH 6.0.
	3. Formulation buffer: 25 mM sodium acetate, pH 4.0.

2.2   Ethanolic Lipid Formulation Solution

1. Cationic or ionizable lipid.
2. 1,2-Distearoyl-sn-glycero-3-phosphocholine (DSPC).
3. Cholesterol.
4. PEG-lipid.
5. Absolute ethanol.

2.3   Microfluidic Manufacturing Apparatus

1. NanoAssemblr™ benchtop instrument (Precision NanoSystems Inc., Vancouver, Canada).
2. NanoAssemblr™ microfluidic cartridge.
3. Disposable syringes: 1–10 mL.

2.4   Post-manufacture Processing

1. 12–14,000 MW cutoff dialysis cassette.
2. PBS (calcium- and magnesium-free).
3. Disposable syringe.
4. 0.2 μm filter.
5. Stir plate and magnetic stir bar.

2.5   siRNA Concentration and Encapsulation Efficiency

1. 96-well plate.
2. siRNA-LNP characterization standard.
3. siRNA-LNP formulations.
4. RiboGreen Assay Kit (Life Technologies, R11490).
5. 1× TE buffer pH 7.5.
6. 1× TE buffer pH 7.5 with 2 % Triton X-100 (Triton buffer).

2.6   Lipid Quantification

1. siRNA-LNP formulations for testing: post-chip, post-dialysis, and post-filtration samples.
2. Wako Cholesterol E (CHOD-DAOS method):
   (a) Cholesterol E standard solution 200 mg/dL (2 mg/mL).
   (b) Cholesterol E color reagent.
   (c) Cholesterol E buffer.

2.7   LNP Characterization Equipment

1. Dynamic light-scattering particle sizer.
2. Fluorescence plate reader.
3. UV/Vis spectrophotometer.

3 Methods

3.1 Microfluidic Manufacture of siRNA-Lipid Nanoparticles (siRNA-LNP)

1. Preparation of siRNA formulation solution: Hydrate lyophilized siRNA in RNA hydration buffer to the desired concentration (generally 10–100 mg/mL). Measure final concentration using A_{260}. Prepare the siRNA formulation solution by diluting the hydrated siRNA into formulation buffer to the desired final concentration (generally 0.1–1.5 mg/mL siRNA). Aliquot 60–80 µL and set aside for siRNA-LNP characterization (*see* Subheadings 3.3–3.5 below).

2. Preparation of lipid formulation solution stock: Solutions of individual lipids are prepared by dissolving a measured amount of dry lipid in absolute ethanol to the desired concentration (generally 10–50 mg/mL lipid, depending on solubility). The lipid formulation solution is prepared by mixing lipid stock solutions to provide a mixture at the appropriate molar ratios and total lipid content (*see* **Note 2**). This mixture is then diluted with absolute ethanol to provide a final solution at the desired volume and lipid concentration. Aliquot 20–30 µL and set aside for siRNA-LNP characterization (*see* Subheadings 3.3–3.5 below).

3. Formulation of siRNA-LNP using the NanoAssemblr™ Benchtop Instrument: Load the NanoAssemblr™ microfluidic cartridge into the NanoAssemblr™ cartridge holder, and select the desired formulation conditions on the NanoAssemblr™ user interface (*see* **Note 3**). Attach a disposable syringe (BD) with formulation buffer into the first inlet of the NanoAssemblr™ microfluidic cartridge. Attach the second syringe containing absolute ethanol into the second inlet of the NanoAssemblr™ microfluidic cartridge. Prime the system by flowing ≥2 mL total of formulation buffer and absolute ethanol through the NanoAssemblr™ microfluidic cartridge using the desired formulation conditions (*see* **Note 4**). Ensure that the syringes are loaded with sufficient volume to run the desired total volume. Once the system has been primed, load the siRNA formulation solution into a disposable syringe of the appropriate size. Remove all air bubbles by gently agitating the syringe. Attach this syringe into the first inlet of the NanoAssemblr™ microfluidic cartridge. Load the lipid formulation solution into a disposable syringe of the appropriate size. Remove all air bubbles by gently agitating the syringe. Attach this syringe into the second inlet of the NanoAssemblr™ microfluidic cartridge. Insert sample collection tube into the left side and a waste collection tube into the right side of the NanoAssemblr™ sample collection block (*see* **Note 5**). Select the desired formulation parameters using the NanoAssemblr™ software interface and click "Go" (*see* **Note 4**). The NanoAssemblr™ Benchtop

Instrument will automate the manufacture of the siRNA-LNP and dispense the formulated nanoparticles into the sample collection tube.

3.2 Post-manufacture Processing of siRNA-LNP

1. Post-manufacture processing: Add PBS to a large beaker at room temperature and stir continuously. The total volume of PBS required is 200–400-fold the recovered sample volume (i.e., if 2 mL of manufactured siRNA-LNP was recovered, 400–800 mL of PBS should be used). Remove dialysis cassette from package and equilibrate in PBS for 10–15 min. Following cassette equilibration, load sample into dialysis cassette, place in beaker of PBS and stir at room temperature for ≥4 h to remove ethanol, and increase sample pH to 7.4. Remove cassette from PBS and recover sample.

2. Sample concentration: If necessary, the recovered sample can be concentrated using a centrifugal filter unit with 10,000 MW cutoff filter. Samples should be spun at 3,000 RPM until the desired concentration is achieved.

3. Sample sterilization: In a sterile hood, filter the recovered sample through a 0.2 μm filter to remove any aggregates and to sterile filter the siRNA-LNP formulation. Set aside approximately 75 μL of the final sample to use for characterization (*see* Subheadings 3.3–3.5 below).

3.3 siRNA-LNP Characterization: Particle Size and Polydispersity

1. Particle size and polydispersity are measured using dynamic light scattering: Dilute sample as needed to achieve an appropriate concentration for sizing (*see* **Note 6**).

3.4 siRNA-LNP Characterization: RNA Encapsulation Efficiency and Concentration (See Note 7)

1. Preparation of sample stock solutions: In the top row of the 96-well plate (Row A, Fig. 2), add 297 μL of TE buffer pH 7.5 to a single well for each sample plus a single well for a PBS blank using a multichannel pipette. Add 3 μL of sample to these wells for a final volume of 300 μL. Add 3 μL of PBS to the blank well. Pipette to mix. This is your stock solution for each sample. The final RNA concentration of these stock solutions should be approximately 4–7 μg/mL.

2. siRNA-LNP sample setup: Add 50 μL of TE buffer pH 7.5 to the two wells directly below each sample (Rows B and C, Fig. 2). Add 50 μL of sample stock solution from Row A into the wells in Rows B and C (this assay is run in duplicate. All liquid handling should be done using a multichannel pipette). Add 50 μL of Triton buffer to the wells in Rows D and E (Fig. 2) below each sample. Add 50 μL of sample stock solution from Row A into the wells in Rows D and E.

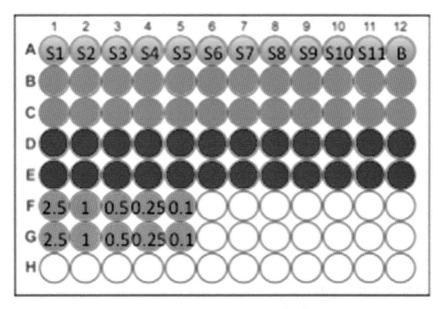

Fig. 2 Experimental design for RiboGreen assay in a 96-well plate. *S1–S11* represent samples *1–11. B* represents blank well. All numerical values represent μL volumes. *Blue wells* = sample; *green wells* = sample + TE buffer pH 7.5; *red wells* = sample + 2 % Triton buffer; *brown wells* = standard curve

3. RiboGreen RNA standard curve setup: Dilute the RNA standard to produce an RNA stock at a final concentration of 20 μg/mL in TE buffer pH 7.5. The final volume should be 150 μL (*see* **Note 8**). Set up a standard curve (in duplicate) as shown in Table 1 using the RNA stock (20 μg/mL siRNA), TE buffer pH 7.5, and Triton buffer. Once samples and standard curve are plated, incubate the plate at 37 °C for 10 min to lyse siRNA-LNP in the presence of Triton X-100 (*see* **Note 9**).

4. Preparation of RiboGreen Solution: Sum the total number of sample wells and standard curve wells. Add three to this number, and multiply the total by 100. This is the total volume, in μL, of RiboGreen Solution needed for this assay. In a 15 mL RNAse-free Falcon tube, dilute the RiboGreen Reagent 1:100 into TE buffer pH 7.5 to the total volume calculated. For example, if 3,000 μL of RiboGreen Solution is needed, add 30 μL of RiboGreen Reagent to 2,970 μL of TE buffer. Vortex the RiboGreen Solution for 10 s to mix.

5. Addition of RiboGreen Solution and sample readings: Remove a 96-well plate from 37 °C incubator. Add 100 μL of RiboGreen Solution to each well. Pop any air bubbles with a needle. Read using fluorescent plate reader (excitation = 480, emission = 525).

6. Sample analysis: Use the data generated by the RiboGreen standard curve to calculate the concentration of siRNA.

Table 1
Preparation of standard curve stock solutions for RiboGreen assay

Final RNA concentration (µg/mL)	RNA stock required (µL)	TE buffer required (µL)	Triton buffer required (µL)	Total volume per well (µL)
2.5	25	25	50	100
1	10	40	50	100
0.5	5	45	50	100
0.25	2.5	47.5	50	100
0.1	1	49	50	100

3.5 siRNA-LNP Characterization: Cholesterol (Lipid) Concentration (See Note 10)

1. Mix the Cholesterol E color reagent and Cholesterol E buffer. This mixture is referred to as "Cholesterol Reagent" in this protocol. There is a separate dry color reagent for each of the two buffers included in the kit. Note the date that color reagent and buffer are mixed and use within 1 month.

2. 96-well plate setup: Sum the total number of Sample Standards and siRNA-LNP formulation samples that will be assayed. Multiply this number by two. This is the number of wells required to run this assay in duplicate (i.e., 12 samples require 24 wells). Add 185 µL of Cholesterol Reagent to the wells (Rows A and B, Fig. 3).

3. For the standard curve (Rows C and D, Fig. 3), add:

   (a) 190 µL of Cholesterol Reagent into two wells.

   (b) 195 µL of Cholesterol Reagent into two wells.

   (c) 197.5 µL of Cholesterol Reagent into two wells.

   (d) 199 µL of Cholesterol Reagent into two wells.

   (e) 185 µL of Cholesterol Reagent into two wells (blank).

4. Sample addition, incubation, and plate reading: Add 15 µL of each sample to the wells from **step 1** (Rows A and B, Fig. 3). This gives a final volume of 200 µL in each sample well. Add 10 µL of Cholesterol E standard solution to wells C1 and D1, 5 µL of Cholesterol E standard solution to wells C2 and D2, 2.5 µL of Cholesterol E standard solution to wells C3 and D3, and 1 µL of Cholesterol E standard solution to wells C4 and D4 (Fig. 4). This gives a final volume of 200 µL in each standard curve well. Add 15 µL of PBS into the two blank wells (wells C5 and D5, Fig. 4). This gives a final volume of 200 µL in each blank well. Place plate at 37 °C for 25 min. Remove plate and read samples at A_{595} using a 96-well plate reader.

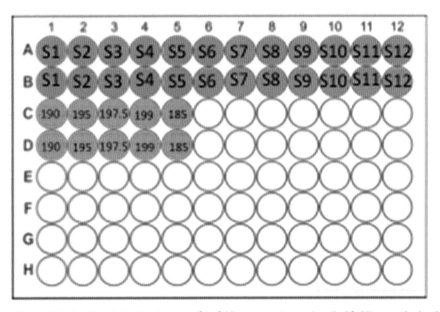

Fig. 3 Experimental design for cholesterol assay. *S1–S12* represent samples *1–12*. All numerical values represent μL volumes of Cholesterol Reagent. *Green wells* = sample; *brown wells* = standard curve

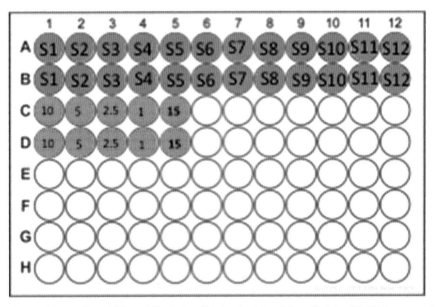

Fig. 4 Experimental layout for addition of Cholesterol E standard to a 96-well plate. *S1–S12* represent samples *1–12*. All numerical values represent μL volumes of Cholesterol E standard. *Green wells* = sample; *brown wells* = standard curve

5. Sample analysis: Use the data generated by the standard curve to calculate the cholesterol concentration, and use the relative proportion of cholesterol in the lipid composition to estimate the total lipid concentration.

4 Notes

1. Prepare all buffers and solutions using RNAse-free water. All disposables should be sterile and RNAse-free. All surfaces, glassware, and labware should be thoroughly cleaned with a 70 % ethanol solution and RNAse Zap (Invitrogen) prior to formulation to prevent RNA degradation.

2. Typical lipid composition for the manufacture of siRNA-LNP is the following: cationic lipid/DSPC/cholesterol/PEG-lipid (50:10:38.5:1.5 mol:mol); siRNA/lipid ratio of 0.06 (wt/wt).

3. Operation of the NanoAssemblr™ Benchtop Instrument is controlled by a custom user interface that enables selection of manufacturing parameters: total formulation volume (range 1–20 mL); syringe size (1, 3, 5, or 10 mL); total flow rate of aqueous siRNA solution and ethanolic lipid solution through the microfluidic cartridge (range 2–12 mL/min); aqueous to ethanol flow rate ratio (1:1–5:1); and sample switching time (start of run, 0–24 s; end of run, 0–6 s). The sample switching function is designed to collect suboptimal particles that are manufactured at the beginning and at the end of the run when the mixing rate is not at steady state. This material is collected in the waste container. For more information, see the NanoAssemblr™ Benchtop Instrument user manual (www.nanoassemblr.com).

4. Typical manufacturing parameter settings on the Nano Assemblr instrument are:
   (a) Total formulation volume = 2 mL.
   (b) Syringe size = 3 and 1 mL.
   (c) Total flow rate = 12 mL/min.
   (d) Aqueous: ethanol flow rate ratio = 3:1.
   (e) Flow rate for aqueous siRNA solution = 9 mL/min.
   (f) Flow rate for ethanolic lipid solution = 4 mL/min.
   (g) Volume of aqueous siRNA solution dispensed = 1.5 mL.
   (h) Volume of ethanolic lipid solution dispensed = 0.5 mL.
   (i) Sample switching time at start of run = 0.25.
   (j) Sample switching time at the end of run = 0.05.

5. Refer to the NanoAssemblr™ Benchtop Instrument user manual for a detailed description of instrument operation (www.nanoassemblr.com).

6. Based on the lipid composition and the NanoAssemblr™ manufacturing settings described in **Note 3** above, particle size is typically 60–65 nm with PDI < 0.05.

7. The RiboGreen assay (Life Technologies) is a fluorescence-based assay for the detection of RNA in solution. This 96-well plate assay is used to determine the encapsulation efficiency and total siRNA concentration of siRNA-LNPs formulated on the NanoAssemblr™. siRNA encapsulation efficiency is determined by measuring fluorescence in the absence or presence of detergent (Triton X-100). Sample Standards (aqueous siRNA and lipids in ethanol mixed by pipette at a ratio equivalent to the flow ratio on the NanoAssemblr™) are used as RNA concentration standards to account for the effect of lipids on RiboGreen fluorescence. Up to 11 samples can be run on a single plate. Please refer to the manufacturer's protocol for more details.

8.

$$V_{RNA\ Stock} = \frac{(20\ \mu g / mL) \times (150\ \mu l)}{C_{RNA\ Stock}}$$

Volume TE buffer = 150 μL − $V_{RNA\ Stock}$

$V_{RNA\ stock}$ = volume of RNA stock required in μL

$C_{RNA\ stock}$ = concentration of RNA stock in μg/mL

9. This assay uses the Sample Standard from siRNA-LNP formulation as the RNA standard. The Sample Standard is prepared by mixing aliquots of the siRNA and lipid stock solutions by pipetting. Aliquots should be mixed at the same ratio as the siRNA-LNP formulation (i.e., 3:1 aqueous/ethanol).

10. The cholesterol assay (Cholesterol E kit, Wako Diagnostics) is an enzymatic colorimetric assay to determine the cholesterol content of siRNA-LNP in the pre-manufacture (Sample Standard), post-manufacture, post-dialysis, and post-filtration steps of the formulation. This is used to calculate lipid losses and dilution during formulation. Up to 12 samples can be run on a single plate. Please refer to the manufacturer's protocol for more details.

Acknowledgements

This work was supported by the National Science and Engineering Research Council of Canada (F09-04486), Canadian Institutes for Health Research (111627), and Genome British Columbia.

References

1. Mozafari MR (2005) Liposomes: an overview of manufacturing techniques. Cell Mol Biol Lett 10:711–719

2. Schwendener RA, Schott H (2010) Liposome formulations of hydrophobic drugs. In: Weissig V (ed) Liposomes methods and protocols, 2nd edn. Humana Press, Totowa, NJ, pp 129–138

3. Mozafari MR (2010) Nanoliposomes: preparation and analysis. In: Weissig V (ed) Liposomes methods and protocols, 2nd edn. Humana Press, Totowa, NJ, pp 29–50

4. Chatterjee S, Banerjee DL (2002) Preparation, isolation, and characterization of liposomes containing natural and synthetic lipids. In: Basu SC, Basu M (eds) Liposome methods and protocols, 1st edn. Humana Press Totowa, NJ, pp 3–16

5. Semple SC, Klimuk SK, Harasym TO et al (2001) Efficient encapsulation of antisense oligonucleotides in lipid vesicles using ionizable aminolipids: formation of novel small multilamellar vesicle structures. Biochim Biophys Acta 1510:152–166

6. Belliveau NM, Huft J, Lin PJ et al (2012) Generation and loading of lipid nanoparticles containing siRNA by microfluidic mixing. Mol Ther Nucleic Acids 1:e37

7. DeMello AJ (2006) Control and detection of chemical reactions in microfluidic systems. Nature 442:394–402

8. Jahn A, Vreeland WN, DeVoe DL et al (2007) Microfluidic directed formation of liposomes of controlled size. Langmuir 23:6289–6293

9. Song H, Chen DL, Ismagilov RF (2006) Reactions in droplets in microfluidic channels. Angew Chem Int Ed Engl 45:7336–7356

10. Song Y, Hormes J, Kumar CS (2008) Microfluidic synthesis of nanomaterials. Small 4:698–711

Chapter 7

Microneedle-Iontophoresis Combinations for Enhanced Transdermal Drug Delivery

Ryan F. Donnelly, Martin J. Garland, and Ahlam Zaid Alkilani

Abstract

It has recently been proposed that the combination of skin barrier impairment using microneedles (MNs) coupled with iontophoresis (ITP) may broaden the range of drugs suitable for transdermal delivery as well as enabling the rate of delivery to be achieved with precise electronic control. However, few reports exist on the combination of ITP with in situ drug-loaded polymeric MN delivery systems. Our in vitro permeation studies revealed that MN enhances transdermal drug delivery. The combination of dissolving MN and ITP did not further enhance the extent of delivery of the low molecular weight drug ibuprofen sodium after short application periods. However, the extent of peptide/protein delivery was significantly enhanced when ITP was used in combination with hydrogel-forming MN arrays. As such, hydrogel-forming MN arrays show promise for the electrically controlled transdermal delivery of biomacromolecules in a simple, one-step approach, though further technical developments will be necessary before patient benefit is realized.

Key words Microneedles, Iontophoresis, Transdermal drug delivery

1 Introduction

The rapid growth of the biotechnology industry has seen an increase in the interest for use of peptide-/protein-based molecules as therapeutic agents [1]. In order for this new generation of therapeutic agents to gain prominence as mainstream treatments of choice, a parallel development of efficient delivery systems will be required by the pharmaceutical industry [2]. Traditionally, peptide and protein drugs have had to be administered parenterally, due to their propensity to undergo acid degradation within the gastrointestinal tract [3]. However, recent advances have seen the development of facilitated technologies that may make transdermal peptide/protein delivery a feasible option. The transdermal route offers a range of advantages for delivery of peptide/protein therapeutics. As proteins have short plasma half-lives, the possibility of continuous delivery offered by the transdermal route is a major benefit, as is the low proteolytic activity in comparison to other routes [1, 2].

Kewal K. Jain (ed.), *Drug Delivery System*, Methods in Molecular Biology, vol. 1141,
DOI 10.1007/978-1-4939-0363-4_7, © Springer Science+Business Media New York 2014

Furthermore, avoidance of first-pass hepatic metabolism eliminates another obstacle to successful systemic peptide/protein delivery. The major challenge associated with delivery of molecules across the skin is overcoming the *stratum corneum* barrier. Thus, a number of enhancement methods have been proposed and investigated to improve bioavailability of peptides/proteins administered across the skin.

The development of microneedle (MN) arrays is likely to play a major role in the progression of the transdermal route as a viable option for the noninvasive systemic delivery of therapeutic peptide/protein agents. MNs are devices composed of micron-size needles which have the ability to physically disrupt the *SC* and create microconduits for drugs to pass through deeper layers of the skin to the microcapillary bed for systemic absorption [2]. MN technology has been demonstrated to be painless [4] and associated with only very minor local adverse skin reactions [5, 6]. As such MN technology appears to be an attractive minimally invasive and patient-friendly drug delivery option for macromolecular therapeutics. MNs have been fabricated in a wide range of designs from various materials, including silicon [7], glass [8], metal [9], and polymers [10–12].

Recently, it has been suggested that, due to the fact that MN application to the skin will result in the creation of aqueous pathways of low electrical resistance [13], the combination of MN technologies and iontophoresis (the application of a small electric current, typically ≤ 0.5 mA/cm^2, to drive ionic and polar molecules across the skin) may lead to a synergistic enhancement in transdermal delivery with the added benefit of precise electronic control [14, 15]. Indeed it has been shown that the combination of MN and ITP can lead to an increased rate of transdermal delivery for a range of drug molecules, both in vitro and in vivo [13, 15–17]. However, to date the combination of ITP and MNs has focused on either the use of nondrug-loaded solid MNs that are used to merely puncture the skin prior to the application of an electrically conducting drug formulation or the use of an electric stimulus to facilitate drug movement through the central bore of hollow MNs. In these cases, the pores created within the skin could potentially close or, in the case of the hollow MNs, the central bore may become blocked by compressed dermal tissue during MN insertion, thus negating the length of benefit that may be obtained through the use of ITP to stimulate drug movement. The use of drug-loaded dissolving polymer MN systems and hydrogel-forming MN may overcome some of the abovementioned issues and thus allow for the in situ combination of MN and ITP in a one-step approach for enhanced transdermal delivery of a wide range of drug molecules. The practical experimental approaches taken to investigate these ITP-MN combinations are described in detail in this chapter.

2   Materials

2.1   Microneedles

Microneedles (MN) were prepared using silicone elastomer micromoulds using methods described previously [12]. Two MN formulations were investigated. The first was a dissolving MN platform formulated from aqueous blends composed of 70 % w/w of a 30 % w/w hydrolyzed gel of poly(methylvinylether-co-maleic acid) (Gantrez® AN-139) neutralized to pH 7 and 30 % w/w of ibuprofen sodium. The second was an integrated hydrogel-forming MN system prepared from aqueous blends of 15 % w/w of hydrolyzed Gantrez® AN-139 cross-linked by esterification with 7.5 % w/w poly(ethylene glycol) 10,000 Da. The latter system serves as a tool to pierce the *stratum corneum* barrier, whereupon it imbibes skin interstitial fluid to form a continuous, unblockable conduit between an attached drug-loaded patch (prepared from aqueous blends of 10 % w/w hydrolyzed Gantrez® AN-139 and 5 % w/w tripropylene glycol monomethyl ether) and the dermal microcirculation.

2.2   Chemicals

Gantrez® AN-139, a copolymer of methyl vinyl ether and maleic anhydride (Mw = 1,080,000), was obtained from Ashland, Guildford, UK. All other chemicals were obtained from Sigma-Aldrich, Dorset, UK, and were of analytical reagent grade.

3   Methods

3.1   Combined Effect of ITP and Integrated MN

MNs have been widely hailed as exciting alternatives to parenteral injection, complete with enhanced ease of use and reduced potential for transmission of infection. However, MN-based delivery typically relies on relatively slow drug diffusion through MN-induced holes in the *stratum corneum* or dissolution of drug-loaded polymeric MN or drug coatings on silicon or metal MN. This is in stark contrast to the rapid delivery possible using conventional needles and syringes. It has recently been proposed that the combination of skin barrier impairment using MN coupled with iontophoresis (ITP) may allow rapid delivery of vaccines or hormones (e.g., insulin) as well as enabling delivery of conventional drug substances to be precisely controlled [13, 16]. Ultimately, this may enable bolus, pulsatile, or responsive drug administration. Combination of MN and ITP has shown to lead to a synergistic enhancement in transdermal delivery of a range of molecules [13, 15–18]. However, the necessity for a two-stage application process (MN applied and removed, then ITP applied) when using this approach may be cumbersome for clinicians and patients, particularly given the importance of ensuring that the drug-loaded formulation is placed correctly over the microporated skin area. Secondly, it is known that the process of skin closure begins almost immediately upon MN

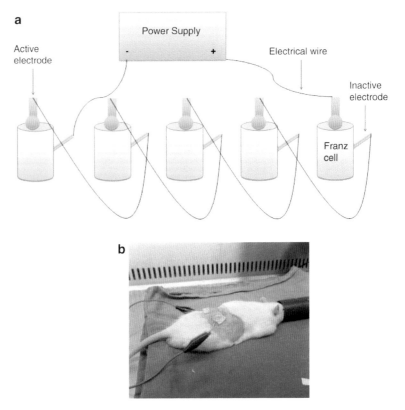

Fig. 1 Schematic illustration of experimental set up employed during in vitro iontophoretic investigations (**a**) and in vivo rat studies (**b**)

removal, unless the skin area is kept under heavy occlusion [19]. This effectively limits the time available for enhanced drug delivery. Such problems may be overcome by the amalgamation of ITP with in situ polymeric MN delivery systems. As such, the present study evaluated the potential for the combination of our novel hydrogel MN devices and ITP. In vitro investigations were initially carried out using the generic setup shown in Fig. 1. Based on the outcomes of these studies, targeted in vivo evaluations were performed using rat models.

Figure 2a illustrates the in vitro transdermal permeation profiles of ibuprofen sodium following cathodal iontophoresis, at a current strength of 0.5 mA and current duration of 6 h, from the dissolving MN platform across excised neonatal porcine skin (350 μm). It was found that the combination of cathodal IP and MN led to enhancement in ibuprofen sodium permeation, in comparison to MN alone, but this was not significant ($p > 0.05$). The cumulative amount of ibuprofen sodium permeated across neonatal porcine skin at 6 h was found to be $26,646 \pm 4,302$ μg and $31,980 \pm 7,746$ μg for MN alone and MN + ITP delivery strategies, respectively. Figure 2b shows the in vitro permeation of ibuprofen

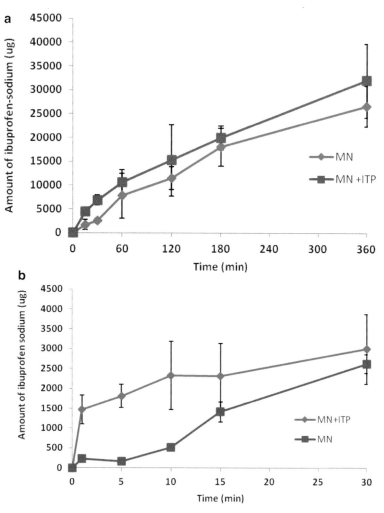

Fig. 2 In vitro permeation of ibuprofen sodium across dermatomed neonatal porcine skin (350 μm) following the combination of an electric current of 0.5 mA for a period of 6 h with dissolving MN (MN + ITP) or passive release from dissolving MN alone (Means ± SD, $n = 5$) (**a**). In vitro permeation of ibuprofen sodium across dermatomed neonatal porcine skin (350 μm) following the combination of an electric current of 0.5 mA for a period of 15 min with dissolving MN (MN + ITP) or passive permeation from dissolving MN alone (Means ± SD, $n = 5$) (**b**)

sodium following the combination of dissolving MN and cathodal ITP for a period of 15 min and passive permeation from dissolving MN alone (MN). The amount of ibuprofen sodium permeating from the combination of MN + ITP was significantly higher ($p < 0.05$) than MN alone at 1, 5, and 10 min. However, after that, there was no significant difference between the two strategies.

Figure 3a–e illustrates the combined effect of ITP and integrated hydrogel MN on permeation of five different solutes across dermatomed neonatal porcine skin. In general, the combination of

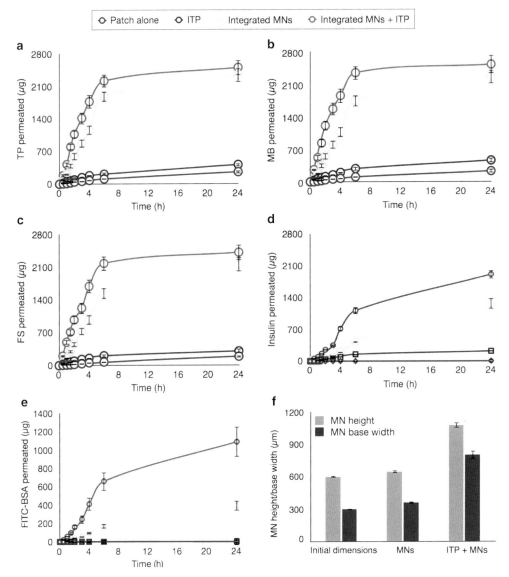

Fig. 3 Combined effect of ITP and integrated hydrogel-forming MN on in vitro permeation of different drug molecules across dermatomed neonatal porcine skin of 300 ± 50 μm thickness and the effect of ITP on in-skin MN swelling. (**a–e**) Show cumulative permeation of TP, MB, FS, insulin, and FITC-BSA, respectively. (**f**) Shows OCT assessment of changes in MN dimensions following insertion into rat skin for a period of 1 h ex vivo. (**g–i**) Representative light microscopy images of hydrogel-forming MN; **g** before skin insertion, **h** 1 h after insertion into skin and **i** 1 h after insertion into skin and application of a continuous electric current of 0.5 mA. (**j**) Shows a representative OCT image of a hydrogel-forming MN array following insertion into rat skin ex vivo and application of an electric current of 0.5 mA for a period of 1 h. (**k**) Shows the % drop in blood glucose levels following application of an electric current (0.5 mA for 2 h) with the insulin-loaded patch and in combination with integrated hydrogel-forming MN. (**l**) Shows plasma concentration of FITC-BSA delivered from integrated MN patches—no detectable FITC-BSA was delivered from the patches alone. (**m**) Shows plasma concentration of FITC-BSA delivered from integrated MN patches combined with ITP. Scale bars represent a length of 300 μm. Error bars indicate standard deviations, $n = 5$ for in vitro studies, $n = 3$ for in vivo studies

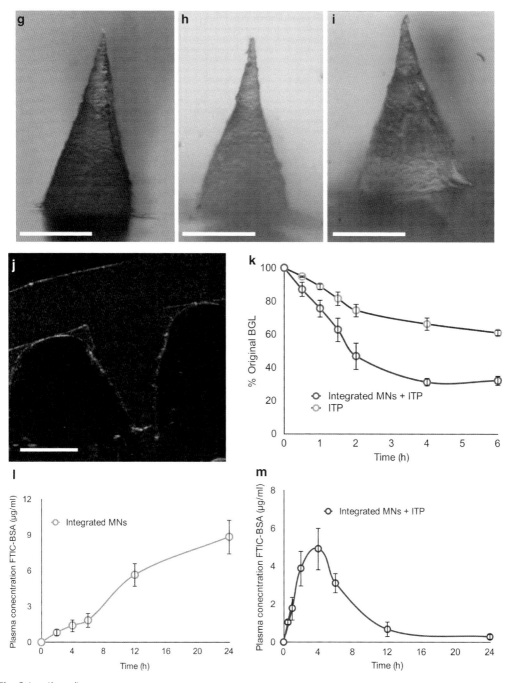

Fig. 3 (continued)

ITP with the novel hydrogel MN array led to a greater rate and overall extent of transdermal delivery, in comparison to integrated MN alone. However, significant enhancements were only observed for the biomolecules insulin (molecular weight approximately 6,000 Da) and fluorescein isothiocyanate-labelled bovine serum albumin (FITC-BSA, molecular weight approximately 66,000 Da). This was particularly notable for FITC-BSA and is understandable, given the approximate 12,000 Da cutoff associated with conventional ITP [18, 20].

In an additional ex vivo experiment, optical coherence tomographic (OCT) analysis revealed that application of an electrical current leads to a dramatic increase in the swelling of the hydrogel-forming MN. It can be seen in Fig. 3f–i that, 1 h after MN insertion into rat skin, MN height increased from ≈610 to 650 μm, while the MN base width increased from ≈307 to 360 μm. However, following application of an electrical current for 1 h, MN height increased to ≈1,076 μm ($p < 0.001$) and the MN base width to approximately 802 μm ($p < 0.001$). This equates to an approximate 5.75-fold increase in MN surface area, in comparison to integrated hydrogel MN swelling under passive conditions. Such pronounced dimensional changes, most likely due to enhanced water uptake by electroosmosis as we have previously postulated for the PEG-cross-linked PMVE/MA system [21], may partially explain the enhancements in delivery observed. The ability of these MNs to imbibe skin interstitial fluid is likely to find use in minimally invasive patient monitoring, since drug concentrations in skin interstitial fluid frequently reflect those in plasma [22]. This would overcome many of the problems associated with direct blood sampling.

3.2 In Vivo Studies

In order to demonstrate the utility and advantages of our novel hydrogel MN technology, we carried out a range of in vivo rat experiments (Fig. 3k–m). Anodal ITP alone (applied for a 2 h period in diabetic rats, after which the electrodes and insulin-loaded adhesive patch were removed) resulted in a maximal drop in the rats' mean blood glucose level (BGL) to ≈61 % of its original value within 6 h, with BGL returning to normal by 12 h. The combination of integrated hydrogel MN and anodal ITP (applied for a 2 h period, after which the electrodes and MN/insulin patch were removed) led to a rapid reduction in BGL, dropping to ≈47 % within 2 h and ≈32 % within 6 h, before returning to normal values by 12 h. It was observed that the combined effect of integrated hydrogel MNs and ITP showed a significantly greater C_{max} (maximum % decrease in BGL value) (70.42 ± 1.86 %) than ITP alone (39.06 ± 2.10 %) ($p < 0.001$) or MN alone (63.16 ± 2.82 %) ($p < 0.001$). Furthermore, the time taken to reach this value was significantly reduced through the combination of integrated MN and ITP (4 h), in comparison to ITP (6 h) ($p < 0.001$) or MN alone (12 h) ($p < 0.001$).

Sustained transdermal delivery of the high molecular weight protein FTIC-BSA was observed following application of integrated MN, with peak plasma concentrations reaching 8.86 ± 1.49 μg/ml at 24 h. The combination of integrated hydrogel MN and ITP led to a significantly accelerated FTIC-BSA permeation, with detectable FTIC-BSA levels found in plasma after only 30 min. Following termination of the electric current and complete removal of the MN device after 2 h of application, FTIC-BSA plasma levels were found to be 3.89 ± 0.96 µg/ml. This is ≈ 4.8-fold greater than the 0.81 ± 0.32 µg/ml detected at the same period through the use of integrated hydrogel MN alone ($p < 0.001$). It was found that the FTIC-BSA plasma levels continued to rise to peak levels of 4.92 ± 1.15 µg/ml at 4 h. In contrast, application of an adhesive FTIC-BSA-loaded patch, or the use of ITP alone, did not result in detection of FTIC-BSA in plasma.

It was again observed that application of an electrical current led to a marked increase in the rate and extent of in-skin swelling of the hydrogel-forming MN arrays. Importantly, as with in vitro studies, the integrated hydrogel MN arrays were removed fully intact following the complete period of application in all animals, regardless of whether ITP was employed or not. This reflects the retention of good mechanical strength by the swollen hydrogel and is likely to be of great importance as MN technology moves forwards towards commercialization, given that regulatory authorities, health-care professionals, and ultimately patients may have concerns about local or systemic reactions that may occur if portions of the needle were to break within the skin. It is our assertion that these integrated hydrogel MNs are well suited for sustained delivery of both small water-soluble drugs and also higher molecular weight peptide and protein molecules. Pulsatile delivery can be achieved through application of ITP to meet on-demand requirements (e.g., delivery of insulin after a meal) or for rapid vaccine delivery.

4   Notes

A number of practical points are important when considering the use of MN/ITP combinations. These are summarized below:

1. MN made from solid silicon or metal needs to be applied and removed prior to applying iontophoretic patches so as to make the additional low resistance pathways created open for movement of charge carriers. Leaving them in place will still allow drug delivery, but this will be reduced by the mechanical obstruction.

2. Micromoulded polymeric MN must have sufficient mechanical strength to penetrate the skin's stratum corneum barrier without breaking. Take care with highly loaded dissolving MN, as drug incorporation can compromise strength.

3. When setting up in vitro Franz cell experiments, equilibrate the skin over the receiver compartment for around 1 h, then remove, dab dry with filter paper and apply the skin to the donor compartment with a cyanoacrylate glue. Take care not to allow the glue to get onto the *stratum corneum*, as this will obstruct MN insertion. Next, with the MN on top of the skin, facing

downwards, in the receiver compartment, use an applicator device to apply the MN. Using a finger is difficult due to the positioning of the MN. Place the donor compartment onto the Franz cell, making sure there are no air bubbles beneath the skin, and clamp the assembly together. Attach the delivery electrode to the MN surface using a gel electrode pad and place the return electrode down the sampling arm. This may need to be momentarily removed during sampling and receiver medium replacement post-sampling.

4. Hydrogel-forming MN will typically only allow current to flow once they have swollen in skin interstitial fluid. This may lead to a lag time of up to 15 min before appreciable transdermal drug delivery is observed. The addition of ITP will, however, allow overall enhanced rates of delivery of macromolecules relative to hydrogel-forming MN alone.

5. In contrast, dissolving MN containing low molecular weight drugs will show an almost instantaneous transdermal delivery of drug, which will be enhanced relative to passive delivery using the MN alone for a short time up to 30 min. Thereafter, ITP will not provide any enhancement in release.

6. ITP can only be applied for relatively short periods of time in vivo. This is due to the need to keep the animals unconscious during current application from a power pack employing wire electrodes. Continuous ITP in vivo may be possible using a miniaturized self-adhesive device with built-in power supply and electrodes.

5 The Future

Microneedles are set to come to major prominence in the coming years due to their ability to deliver hydrophilic drug molecules, particularly those of high molecular weight across the skin's *stratum corneum* barrier. It is obvious that extensive safety studies will be required by regulatory authorities, given that the protective barrier of the skin is being compromised. Combining microneedles with iontophoresis may be useful when rapid or pulsatile delivery is required. However, such applications may only be truly beneficial when macromolecules are to be delivered since, as we have shown here and previously, small molecules are already delivered efficiently using microneedles alone, with iontophoresis only eliciting enhanced drug administration at very short application times.

Researchers in academia and industry will have to give considerable thought to the design of combined microneedle-iontophoresis devices, given the lack of success in commercialization of straightforward iontophoresis-only products. The overly complex, bulky, and expensive nature of such systems has greatly impaired patient acceptance and contributed to prescriber reticence. One idea might be to design a custom application device that would insert the microneedles into the skin and then apply the electrical current for a short period of time, say, 1–2 min, to mimic a conventional needle- and syringe-based injection. If such an applicator could also measure skin impedance, successful microneedle insertion, which immediately reduces skin impedance by opening up low resistance pathways in the *stratum corneum*, could be confirmed to the patient/clinician. Furthermore, if the applicator could also follow the reduction in conductivity of a microneedle patch associated with delivery of the charge carrying drug into the skin, then drug dosing could be accurately controlled. The technology to make such a device is already available, though never before applied in this way. Further advances in this field have the potential to make a meaningful difference to patient health, particularly in needle-phobic patients and those in the developing world requiring safe vaccination.

References

1. Prausnitz MR (2004) Microneedles for transdermal drug delivery. Adv Drug Deliv Rev 56(5):581–587
2. Banga A (2007) Transdermal delivery of proteins. Pharm Res 24:1357–1359
3. Migalska K, Morrow DI, Garland MJ, Thakur R, Woolfson AD, Donnelly RF (2011) Laser-engineered dissolving microneedle arrays for transdermal macromolecular drug delivery. Pharm Res 28(8):1919–1930
4. Kaushik S, Allen H, Donald D et al (2001) Lack of pain associated with microfabricated microneedles. Anesth Analg 92:5024
5. Bal SM, Caussin J, Pavel S, Bouwstra JA (2008) In vivo assessment of safety of microneedle arrays in human skin. Eur J Pharm Sci 35(3): 193–202
6. Van Damme P, Oosterhuis-Kafeja F, Van der Wielen M et al (2009) Safety and efficacy of a novel microneedle device for dose sparing intradermal influenza vaccination in healthy adults. Vaccine 27(3):454–459
7. Wilke N, Mulcahy A, Ye S, Morrissey A (2005) Process optimization and characterization of silicon microneedles fabricated by wet etch technology. Microelectron J 36:650–656
8. McAllister DV, Wang PM, Davis SP et al (2003) Microfabricated needles for transdermal delivery of macromolecules and nanoparticles: fabrication methods and transport studies. Proc Natl Acad Sci U S A 100(24):13755–13760
9. Cormier M, Johnson B, Ameri M et al (2004) Transdermal delivery of desmopressin using a coated microneedle array patch system. J Control Release 97(3):503–511
10. Sullivan SP, Murthy N, Prausnitz MR (2008) Minimally invasive protein delivery with rapidly dissolving polymer microneedles. Adv Mater 20(5):933–938
11. Chu LY, Choi SO, Prausnitz MR (2010) Fabrication of dissolving polymer microneedles for controlled drug encapsulation and delivery: bubble and pedestal microneedle designs. J Pharm Sci 99(10):4228–4238
12. Donnelly RF, Majithiya R, Singh TR et al (2011) Design, optimization and characterisation of polymeric microneedle arrays prepared by a novel laser-based micromoulding technique. Pharm Res 28(1):41–57
13. Lanke SS, Kolli CS, Strom JG, Banga AK (2009) Enhanced transdermal delivery of low molecular weight heparin by barrier perturbation. Int J Pharm 365(1–2):26–33
14. Wang Y, Thakur R, Fan Q, Michniak B (2005) Transdermal iontophoresis: combination strategies to improve transdermal iontophoretic drug delivery. Eur J Pharm Biopharm 60(2):179–191
15. Katikaneni S, Badkar A, Nema S, Banga AK (2009) Molecular charge mediated transport of a 13 kD protein across microporated skin. Int J Pharm 378(1–2):93–100
16. Wu XM, Todo H, Sugibayashi K (2007) Enhancement of skin permeation of high molecular compounds by a combination of microneedle pretreatment and iontophoresis. J Control Release 118(2):189–195

17. Chen H, Zhu H, Zheng J et al (2009) Iontophoresis-driven penetration of nanovesicles through microneedle-induced skin microchannels for enhancing transdermal delivery of insulin. J Control Release 139(1):63–72

18. Vemulapalli V, Yang Y, Friden PM, Banga AK (2008) Synergistic effect of iontophoresis and soluble microneedles for transdermal delivery of methotrexate. J Pharm Pharmacol 60(1): 27–33

19. Kalluri H, Banga AK (2011) Formation and closure of microchannels in skin following microporation. Pharm Res 28(1):82–94

20. Williams AC (2003) Transdermal and topical drug delivery. Pharmaceutical Press, London

21. Garland MJ, Singh TR, Woolfson AD, Donnelly RF (2011) Electrically enhanced solute permeation across poly(ethylene glycol)-crosslinked poly(methyl vinyl ether-co-maleic acid) hydrogels: effect of hydrogel crosslink density and ionic conductivity. Int J Pharm 406(1–2): 91–98

22. Mukerjee EV, Collins SD, Isseroff RR, Smith RL (2004) Microneedle array for transdermal biological fluid extraction and in situ analysis. Sens Actuat A Phys 114:267–275

Chapter 8

Polymer Nanoparticle-Based Controlled Pulmonary Drug Delivery

Moritz Beck-Broichsitter, Alexandra C. Dalla-Bona, Thomas Kissel, Werner Seeger, and Thomas Schmehl

Abstract

The development of novel formulations for controlled pulmonary drug delivery purposes has gained remarkable interest in medicine. Although nanomedicine represents attractive concepts for the treatment of numerous systemic diseases, scant information is available on the controlled drug release characteristics of colloidal formulations following lung administration, which might be attributed to the lack of methods to follow their absorption and distribution behavior in the pulmonary environment.

In this chapter, we describe the methods of preparation and characterization of drug-loaded polymeric nanoparticles prepared from biodegradable charge-modified branched polyesters, aerosolization of the nanosuspensions using a vibrating-mesh nebulizer, and evaluation of the pulmonary pharmacokinetics (i.e., absorption and distribution characteristics) of the nanoscale drug delivery vehicles following aerosol delivery to the airspace of an isolated lung model. The disclosed methodology may contribute to the design of advanced colloids for the treatment of respiratory disorders.

Key words Biodegradable charge-modified branched polyesters, Polymeric nanoparticles, Controlled pulmonary drug delivery, Vibrating-mesh nebulization, Isolated lung model

1 Introduction

Inhalative delivery of therapeutic aerosols is a typical example of a targeted treatment of lung disorders, which offers a number of advantages over other routes of administration, e.g., pulmonary selectivity and reduced side effects [1, 2]. Clearly, the advantages of aerosol therapy depend on the rate and mechanisms of elimination of the delivered medication at the target site. The prompt decay of the drug in the lung often necessitates frequent daily inhalations, conflicting with the patients' convenience and compliance [3].

To overcome the disadvantages of conventional inhalation therapy, pulmonary delivery systems with controlled release properties such as liposomes and micro- as well as nano-sized particles have attracted growing attention [4, 5]. Among them, polymeric

Kewal K. Jain (ed.), *Drug Delivery System*, Methods in Molecular Biology, vol. 1141,
DOI 10.1007/978-1-4939-0363-4_8, © Springer Science+Business Media New York 2014

nanoparticles that bypass lung clearance and thus control the residence time of the encapsulated drug within the respiratory tract represent attractive candidates [6–8]. Moreover, polymeric nanoparticles are thought to fulfill the stringent requirements of delivery systems intended for pulmonary application such as sufficient drug loading [9], controlled release of the therapeutic agent [10, 11], stability upon aerosolization [10–16], and low toxicity [17].

As the biological environment impacts the performance of pharmaceutical formulations, conventional in vitro studies may not reflect the delivery situation under in vivo conditions impeding the design of advanced drug delivery vehicles for lung application [18]. However, ex vivo models were generally used to examine lung-specific absorption and distribution characteristics of inhaled medications [19–21].

Here, we describe the preparation and characterization of promising drug-loaded nanoparticles synthesized from novel biodegradable charge-modified branched polyesters. Vibrating-mesh technology and adjustment of the nanoparticle preparation process facilitate the formation of aerosol clouds composed of unaffected nanosuspensions. Moreover, drug-loaded polymeric nanoparticles are transferrable to the airspace of an isolated lung model, which allows investigation of lung-specific behavior of the inhaled formulation.

2 Materials

2.1 Dye Solution

1. Isotonic phosphate-buffered saline (PBS) is prepared by weighing 8.0 g NaCl, 0.2 g KCl, 1.2 g Na_2HPO_4, and 2.0 g KH_2PO_4 in a beaker and dissolving the solids in 1,000 ml of filtrated, double-distilled water (conductance of 0.055 μS/cm at 25 °C). The pH is adjusted to a value of 7.4 using concentrated HCl and NaOH solution, respectively. The resulting buffer is stored at a temperature of 4 °C.

2. Model drug: 5(6)-carboxyfluorescein (CF) (≥95 %, Sigma-Aldrich) is stored in the dark at room temperature.

3. The fluorescent dye solution is prepared in PBS pH 7.4 (final concentration: 50 μg/ml). The resulting solution is stored in the dark at a temperature of 4 °C.

2.2 Nanoparticle Preparation

1. Nanoparticle matrix material: poly[vinyl 3-(diethyl amino) propylcarbamate-*co*-vinyl acetate-*co*-vinyl alcohol]-*graft*-*poly*(D,L-lactide-*co*-glycolide) (Fig. 1) abbreviated as DEAPA(39)-PVA-*g*-PLGA (1:10) is stored at –20 °C with a desiccant (e.g., blue silica gel) (*see* **Note 1**).

2. Model drug: CF.

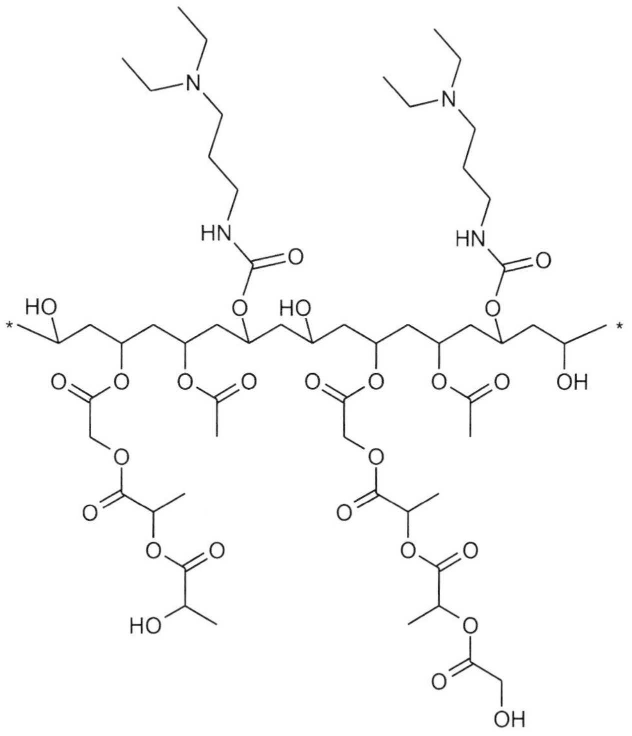

Fig. 1 General structure of DEAPA-PVA-*g*-PLGA. Reproduced with permission from [10]. Copyright 2009 Elsevier

3. Stabilizer: carboxymethyl cellulose (CMC) (Tylopur® C 600, Clariant).

4. Solvents: acetone p.a. (Acros), filtrated, double-distilled water.

5. Multistage magnetic-stirrer plate (e.g., RO 15, IKA®-Werke), electronically adjustable single-suction pump (e.g., PD 5001, Heidolph), injection needle (0.6×30 mm) (e.g., Sterican®, B. Braun), and rotary evaporator (e.g., Rotavapor® RII, Büchi).

2.3 Nanoparticle Characterization

1. Nanoparticle size and ζ-potential: Zetasizer Nano ZS/ ZEN3600, disposable folded capillary cells (DTS 1060) (both from Malvern Instruments).

2. Nanoparticle morphology: NanoWizard® (JPK Instruments), Si_3N_4 tips attached to I-type cantilevers with a length of 230 μm and a nominal force constant of 40 N/m (NSC16 AlBS, Mikromasch).

3. Drug entrapment efficiency: Airfuge®, 200 μl polyallomer tubes (both from Beckman Coulter).

4. In vitro drug release: PBS pH 7.4, Alveofact® (Lyomark Pharma), Rotatherm® (Gebr. Liebisch).

2.4 Nebulization	1. Syringe filters: 5.0 μm, Cameo 30 N (GE, Water & Process Technologies).
	2. Nebulizer: Aeroneb® Pro (Aerogen).
	3. Balance: BP 211 D (Sartorius).
	4. Laser diffraction: HELOS (Sympatec).
2.5 Isolated Lung Model	1. Male, pathogen-free "New Zealand white" rabbits as organ donors (body weight: 2.5–3.5 kg).
	2. Stainless-steel surgical instruments, syringes, catheter, cannula, and tape.
	3. Anesthesia and anticoagulation: ketamine (Ketanest®, Pharmacia), xylazine (Rompun®, Bayer Vital), lidocaine (Xylocaine®, AstraZeneca), heparin (heparin-sodium-5000 ratiopharm®, ratiopharm), and isotonic NaCl solution (B. Braun).
	4. Peristaltic pump (e.g., roller pump BP 742, Fresenius), double-walled glass containers, diverse filters (e.g., Pall Cardioplegia Plus 0.2 μm, Pall) for the circulating perfusion medium and respiration gas, tubing (e.g., C-FLEX®, Cole-Parmer), bubble chamber, thermostat (e.g., F12-ED, Julabo), catheters with pressure sensors connected to pressure transducers (e.g., Combitrans, B. Braun), software (e.g., Labtech Notebook, OMEGA Engineering), temperature-equilibrated housing chamber, force transducer (e.g., Wägezelle Typ U1, Hottinger Baldwin Messtechnik), and respirator (cat/rabbit ventilator, Hugo Sachs Elektronik).
	5. Perfusion medium: Krebs-Henseleit buffer and Na_2CO_3 solution.
	6. Ventilation gas mixture: 5 % CO_2, 16 % O_2, and 79 % N_2.
2.6 Pulmonary Drug Absorption and Distribution Characteristics in the Isolated Lung Model	1. Nebulizer: Aeroneb® Pro including T-shaped mouth piece and tubing adapters.
	2. Pulmonary aerosol deposition: ^{99m}Tc (10–14 MBq), isotonic NaCl solution, and gamma counter (Raytest).
	3. Bronchoalveolar lavage: 50 ml syringes with tracheal cannula and isotonic NaCl solution.
2.7 CF Quantification by Fluorescence Spectroscopy	1. Model drug: CF.
	2. Buffer solution: PBS pH 7.4.
	3. Centrifuge: centrifuge 5418 (Eppendorf).
	4. Fluorescence spectroscopy: LS 50 B (Perkin Elmer), 96-well plates (Nunc).

3 Methods

Numerous methods are available for the fabrication of nanoparticles from preformed polymers [22]. The nanoprecipitation technique is a well-known method for the production of polymeric nanoparticles [23]. Among the suitable polymers, those most frequently employed belong to the family of biocompatible and biodegradable polyesters [24]. However, linear polyesters (e.g., PLGA) reveal a number of limitations as nanoparticle matrix material (e.g., slow degradation rates, low affinity to hydrophilic drugs). Meanwhile, fast degrading polymers (e.g., DEAPA-PVA-*g*-PLGA) have been developed for lung delivery [10, 25, 26]. These novel biodegradable charge-modified branched polyesters promote electrostatic interactions with therapeutic agents of opposite charge, which significantly improves the drug loading of the nanoparticulate carriers [9–11].

3.1 Preparation of Nanoparticles

1. Dissolve DEAPA(39)-PVA-*g*-PLGA (1:10) in acetone for ~12 h at room temperature to yield a stock solution concentration of 5 mg/ml.

2. In parallel, dissolve CF in acetone to add up to a final concentration of 500 µg/ml.

3. The non-solvent phase consists of filtrated, double-distilled water containing 50 µg CMC/mg of DEAPA(39)-PVA-*g*-PLGA (1:10) (*see* **Note 2**).

4. Mix 1 ml of the polymer stock solution with 0.5 ml of the organic fluorescent dye solution for ~1 h at room temperature before starting the nanoprecipitation process.

5. Transfer 5 ml of the aqueous phase into 10 ml scintillation vials equipped with 10×2 mm stirrer bars located on a multi-magnetic-stirrer plate.

6. Inject the organic phase (1.5 ml) at a constant flow rate of 10.0 ml/min into the magnetically stirring (500 rpm) aqueous phase.

7. Remove the organic solvent from the nanosuspension by rotary evaporation (~1 h, ~200 mbar, 25 °C). Replace evaporated water.

3.2 Nanoparticle Characterization

3.2.1 Size and ζ-Potential Measurements

1. The size and size distribution of drug-loaded polymeric nanoparticles are measured by photon correlation spectroscopy (PCS) at 25 °C using noninvasive back scatter technology (scattering angle of 173°) at a sample concentration of ≤1 mg/ml (*see* **Note 3**).

2. The ζ-potential of the samples is analyzed by laser Doppler velocimetry combined with phase analysis light scattering in 1.0 mM NaCl (*see* **Note 4**).

3.2.2 Nanoparticle Morphology

The morphology of drug-loaded polymeric nanoparticles is analyzed by atomic force microscopy (AFM). 10 µl of sample solution is placed on commercial glass slides. Following 10 min of incubation, slides are rinsed with filtered, double-distilled water. Samples are then dried in a stream of dry nitrogen. Measurements are performed in intermittent contact mode at a scan frequency of 0.5–1.0 Hz and antiproportional to the scan size (*see* **Note 5**).

3.2.3 Determination of Entrapped Model Drug

1. Subject 175 µl of freshly prepared nanosuspension to ultracentrifugation (i.e., $199,000 \times g$) for 20 min at 4 °C.

2. Separate an aliquot from the supernatant containing the non-encapsulated CF amount.

3. Quantify the CF concentration in the supernatant by fluorescence spectroscopy (*see* Subheading 3.5) (*see* **Note 6**).

3.2.4 In Vitro Release Studies

The in vitro drug release is studied in PBS pH 7.4 and Alveofact®-supplemented PBS pH 7.4 (0.5 mg/ml). Drug-loaded polymeric nanoparticles containing a theoretical drug loading of 5 % (w/w) are employed for the analysis:

1. Transfer samples (0.4 ml) to 15 ml plastic tubes and dilute them with release medium (final volume: 10 ml).

2. Incubate the samples at 37 °C with vertical rotation (20 rpm).

3. Take 175 µl samples at the predetermined time points (e.g., 10, 20, 30, 45, 60, 90, 120, 180, and 240 min) and subject them to ultracentrifugation (i.e., $199,000 \times g$) for 20 min at 4 °C.

4. Separate an aliquot from the supernatant containing the released CF fraction.

5. Determine the cumulative release of model drug by fluorescence spectroscopy (*see* Subheading 3.5) (*see* **Note 7**).

3.3 Nebulization

Individual drug-loaded polymeric nanoparticles are expected to be exhaled after inhalation owing to their unique size characteristics [27]. Hence, to achieve peripheral lung deposition of said drug delivery devices, application forms for inhalation need to display defined aerodynamic characteristics (i.e., mass median aerodynamic diameter (MMAD) of 1–5 µm). Besides dry powder aerosolization of nanoparticle-containing microparticles [28], nebulization of aqueous nanosuspensions has proven suitable for pulmonary application of nanoparticle formulations [10, 11]. However, a number of polymeric nanosuspensions were prone to aggregation and concentration during nebulization, emphasizing that formulations should only be used in combination with a specific nebulizer [12, 13, 16].

3.3.1 Aerosol Output Rate

1. Fill the nebulizer with a defined amount of filtrated formulation (≥3 ml), weigh the filled device, nebulize the formulation continuously for at least 3 min, and then weigh the device again.

2. Determine the total aerosol output by comparing the device weight before and after nebulization.

3. Calculate the aerosol output rate using the resulting weight difference in g/min.

3.3.2 Laser Diffraction

1. Fill the nebulizer with filtrated formulation (~3 ml), clamp the device into a stand, and position the mouthpiece exit of the nebulizer ~40 mm from the Fourier lens face and ~30 mm from the laser beam axis (if necessary, pass the delivered aerosol through the mouthpiece using an additional gas flow rate of ~5–10 l/min) (*see* **Note 8**).

2. Nebulize the formulation through the laser beam ($\lambda = 632.8$ nm). Ensure an optical obstruction concentration of >10 %.

3. Perform the measurements with six runs of 100×50 ms duration each (lens system: R2 0.25/0.45–87.5 μm).

4. Analyze the diffraction pattern in Mie mode to calculate the volume median diameter (VMD) and geometric standard deviation (GSD) of the generated aerosol clouds (*see* **Note 9**).

5. Nebulize isotonic NaCl solution as a control.

3.3.3 Nanoparticle Stability to Nebulization

1. Nebulize at least 3 ml of filtrated nanosuspension through the T-shaped mouthpiece (additional airflow rate of ~5–10 l/min).

2. Collect the nebulized formulation by impacting the delivered aerosol on commercial glass slides, which are located in front of the mouthpiece.

3. Investigate the characteristics of the collected nanosuspension using PCS, ζ-potential, AFM, and fluorescence spectroscopy measurements (*see* **Note 10**).

3.4 Isolated, Perfused, and Ventilated Lung Model (IPL)

Ex vivo lung models are frequently employed to account for drug absorption profiles after pulmonary challenge [18–21]. Consequently, gained characteristics about the fate of the delivered formulation in the biological environment allow for the design of advanced controlled drug release vehicles to the lung [18].

3.4.1 Surgical Procedure for IPL Preparation

This requires a surgical procedure on a rabbit (*see* **Note 11**) [18, 32]:

1. Fill the experimental system with the perfusion medium and cool the setup to 4 °C (*see* **Note 12**).

2. Anesthetize and anticoagulate the rabbit through the ear vein (Ketanest®/Rompun® ratio of 3/2 (4.5 ml in total), heparin: 1,000 U/kg).

3. After achieving a deep anesthetization, fix the animal on his back. Pull out the tongue to avoid suffocation.

4. Inject xylocaine solution (~8 ml) s.c. in the pretracheal region.

5. Expose the trachea by blunt dissection, and insert and fix a cannula (diameter: 3 mm) in the trachea. Subsequently, ventilate the animal with room air (tidal volume: 10 ml/kg; frequency: 30 strokes/min).

6. After mid-sternal thoracotomy, spread the ribs, incise the right ventricle, and immediately place a fluid-filled perfusion catheter into the pulmonary artery, which needs to be secured with a ligature.

7. After insertion of the catheter, start the perfusion with the cooled perfusion medium (10–20 ml/min), and then cut the heart open at the apex.

8. Excise the trachea, lungs, and heart en bloc from the thoracic cage.

9. Introduce a second catheter with a bent cannula (diameter: 4 mm) via the left ventricle into the left atrium and fix it in this position (*see* **Note 13**).

10. Place the isolated organs in a temperature-equilibrated housing chamber, freely suspended from a force transducer for continuous monitoring of the organ weight.

11. After rinsing the lungs with at least 1,000 ml of perfusion medium for washout of blood, close the perfusion circuit for recirculation. Meanwhile, increase the flow rate slowly to 100 ml/min. At the same time, elevate the temperature of the perfusion medium and housing chamber to 40 °C.

12. In parallel, change to the ventilation gas mixture (positive end-expiratory pressure of $1 \text{ cmH}_2\text{O}$) (*see* **Note 14**) to maintain the pH of the perfusion fluid in the range between 7.35 and 7.45.

13. After a 30 min steady-state period, exchange the perfusion fluid once by fresh perfusate (volume: 300 ml). Adjust the pH of the perfusion medium (*see* **Note 15**).

14. Adjust the left atrial pressure to 1.5 mmHg by changing the hydrostatic pressure caused by the height of the venous part of the system.

15. Monitor the pH of the perfusion medium, pressures in the pulmonary artery, left atrium, and trachea, as well as organ weight throughout the experiment.

3.4.2 Pulmonary Absorption and Distribution Characteristics of Nebulized Formulations in the IPL

For the analysis of the pulmonary pharmacokinetics of the drug-loaded nanoparticles, formulations need to be delivered to the lung model via the intratracheal route by nebulization (Fig. 2) [10, 18]:

1. Connect the nebulizer to the inspiratory tubing between the ventilator and the isolated organ.

2. Nebulize 3 ml of filtrated formulation containing a CF concentration of 50 µg/ml into the IPL.

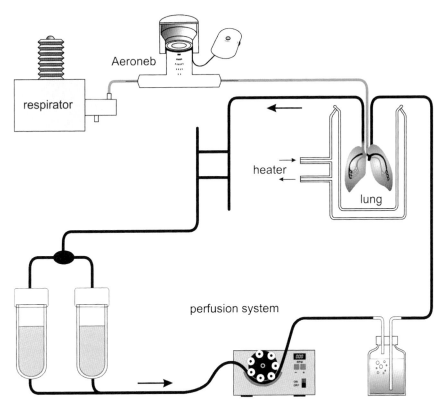

Fig. 2 Schematic depiction of the rabbit lung model useful for pharmacokinetic studies of drug-loaded polymeric nanoparticle formulations. Reproduced with permission from [10]. Copyright 2009 Elsevier

3. Take 800 μl samples from the venous part of the system (e.g., 10, 20, 30, 45, 60, 90, 120, 180, and 240 min after nebulization).

4. Perform a bronchoalveolar lavage at the end of the absorption experiment. Instill and reaspirate three times 50 ml of isotonic NaCl solution through the trachea.

5. Centrifuge all samples at $300 \times g$ for 10 min to remove cells.

6. Determine the sample concentrations by fluorescence spectroscopy (*see* Subheading 3.5) (*see* **Note 16**).

    To determine the percentage of model drug absorbed into the perfusate and distributed within the IPL, the deposited amount of aerosol needs to be determined in separate experiments [10].

7. Connect the nebulizer to the IPL model as described above.

8. Dissolve ^{99m}Tc in 10 ml of isotonic NaCl solution.

9. Nebulize 3 ml of this solution into the IPL.

10. Clamp the lung and stop perfusion during aerosol delivery.

11. Determine the radioactivity of the expiratory filter (exhaled fraction) and lung (lung deposition) by gamma counting (*see* **Note 17**).

3.5 CF Quantification by Fluorescence Spectroscopy

1. Dilute the CF stock solution (50 µg/ml) with PBS pH 7.4 to reach a range of CF concentrations between 5 and 50 ng/ml.

2. Dilute samples with PBS pH 7.4 if necessary.

3. Measure 200 µl of each sample for its fluorescence intensity (*see* **Note 18**).

4. Calculate the CF content using a calibration curve ($R^2 > 0.995$).

4 Notes

1. Biodegradable charge-modified branched polyesters were synthesized and characterized as described in detail elsewhere [26]. These polymers comprise short poly(D,L-lactide-*co*-glycolide) (PLGA) chains grafted onto an amine-substituted poly(vinyl alcohol) (PVA) backbone. As abbreviation A-PVA(x)-g-PLGA (1:y) is used. A is the abbreviation of the type of amine substitution (DEAPA for 3-(diethylamino)pro-pylamine), x represents the total average number of amine functions on the PVA backbone, and y is the PLGA side-chain length [10]. The employed polyesters are hygroscopic and undergo rapid degradation upon contact with water. The packaged polymer samples need to reach room temperature before opening in order to minimize water uptake. Optimally, poly-mers are handled under a glove box in a water-free, inert gas atmosphere.

2. CMC acts as a colloidal stabilizer during nanoparticle prepara-tion as well as nebulization [10, 13, 29].

3. Dilution of nanosuspensions to a final nanoparticle concentra-tion of ≤1 mg/ml is important to avoid multiscattering events.

4. A constant ionic strength improves ζ-potential measurements.

5. AFM analysis in intermittent contact mode minimizes damage of the sample surface [30].

6. The CF entrapment efficiency is calculated indirectly by deter-mining the non-encapsulated CF amount.

7. Nanosuspension without CF and pure model drug should be incubated under the same conditions.

8. Positioning of the nebulizer to the instrument avoids vignett-ing (loss of light scattered at large angles) and mouthpiece interference with the expanded laser beam [31]. Moreover, air extraction behind the laser beam ensures that the aerosol does not reenter the laser sensing zone.

9. The optical properties of the samples are as follows for data anal-ysis: real part of the refractive index, 1.33; complex part of the refractive index (i.e., absorption), 0. The value obtained for

the VMD of produced aerosol droplets might be converted to the MMAD using the sample density ($MMAD = VMD \times (\rho_p/\rho_w)^{1/2}$). The GSD is calculated from the laser diffraction values ($GSD = d_{84\%}/d_{16\%}^{1/2}$).

10. Consider the drying effect of the aerosol during collection for determination of the model drug entrapment efficiency.

11. Only lungs that have a homogeneous white appearance with no signs of hemostasis, edema, or atelectasis, a constant mean pulmonary artery and peak ventilation pressure in the normal range (i.e., 4–10 and 5–8 mmHg, respectively), and are iso-gravimetric during an initial steady-state period are considered for experiments [18, 32].

12. Take care that no air bubbles are introduced into the system.

13. The system should not reveal any leakage or obstructions.

14. "Positive" pressure ventilation of the IPL allows for a homogeneous and efficient aerosol deposition within the lung.

15. Artificial perfusion medium (e.g., Krebs-Henseleit buffer) is only adequate for hydrophilic drugs. Presence of albumin relieves the analysis of hydrophobic drug substances [18].

16. In order to consider loss of perfusate during the IPL experiments (~10–15 ml/h), the measured CF perfusate concentration needs to be corrected using the following formula [10, 11]:

$$c_{corr}(t) = \frac{c(t) \cdot V_p(t) + \left[V_p(0) - V_p(t)\right] \cdot \dfrac{c(t)}{2}}{V_p(0)} = \frac{c(t)}{2} \cdot \left[\frac{V_p(t)}{V_p(0)} + 1\right]$$

where $c_{corr}(t)$ is the corrected CF concentration in perfusate after time t, $c(t)$ is the measured CF concentration in perfusate after time t, $V_p(t)$ is the perfusate volume after time t, and $V_p(0)$ is the perfusate volume at the beginning of the experiment.

17. Calculate the deposition fraction (DF) as follows: $DF = LD/IF = LD/(LD + EF)$, where LD is the lung deposition, EF is the exhaled fraction, and IF is the inhaled fraction = $LD + EF$ [10].

18. Fluorescence intensity of samples is measured at $\lambda_{ex} = 490$ nm (slit: 5.0 nm) and $\lambda_{em} = 520$ nm (slit: 5.0 nm) for 1 s.

Acknowledgements

This work was supported by grants from the "Deutsche Forschungsgemeinschaft" (BE 5308/1-1), "Universitätsklinikum Giessen und Marburg" (1/2012 GI), and "Wirtschafts- und Infrastrukturbank Hessen" (*Nanosurfact*).

References

1. Courrier HM, Butz N, Vandamme TF (2002) Pulmonary drug delivery systems: recent developments and prospects. Crit Rev Ther Drug Carrier Syst 19:425–498

2. Groneberg DA, Witt C, Wagner U et al (2003) Fundamentals of pulmonary drug delivery. Respir Med 97:382–387

3. Gessler T, Seeger W, Schmehl T (2011) The potential for inhaled treprostinil in the treatment of pulmonary arterial hypertension. Ther Adv Respir Dis 5:195–206

4. Gaspar MM, Bakowsky U, Ehrhardt C (2008) Inhaled liposomes – current strategies and future challenges. J Biomed Nanotechnol 4:245–257

5. Kurmi BD, Kayat J, Gajbhiye V et al (2010) Micro- and nanocarrier-mediated lung targeting. Expert Opin Drug Deliv 7:781–794

6. Azarmi S, Roa WH, Löbenberg R (2008) Targeted delivery of nanoparticles for the treatment of lung diseases. Adv Drug Delivery Rev 60:863–875

7. Lebhardt T, Roesler S, Beck-Broichsitter M et al (2010) Polymeric nanocarriers for drug delivery to the lung. J Drug Delivery Sci Technol 20:171–180

8. Beck-Broichsitter M, Merkel OM, Kissel T (2012) Controlled pulmonary drug and gene delivery using polymeric nano-carriers. J Contr Release 161:214–224

9. Beck-Broichsitter M, Schmehl T, Gessler T et al (2012) Development of a biodegradable nanoparticle platform for sildenafil: formulation optimization by factorial design analysis combined with application of charge-modified branched polyesters. J Contr Release 157:469–477

10. Beck-Broichsitter M, Gauss J, Packhaeuser CB et al (2009) Pulmonary drug delivery with aerosolizable nanoparticles in an ex vivo lung model. Int J Pharm 367:169–178

11. Beck-Broichsitter M, Gauss J, Gessler T et al (2010) Pulmonary targeting with biodegradable salbutamol-loaded nanoparticles. J Aerosol Med 23:47–57

12. Dailey LA, Schmehl T, Gessler T et al (2003) Nebulization of biodegradable nanoparticles: impact of nebulizer technology and nanoparticle characteristics on aerosol features. J Contr Release 86:131–144

13. Dailey LA, Kleemann E, Wittmar M et al (2003) Surfactant-free, biodegradable nanoparticles for aerosol therapy based on branched polyesters, DEAPA-PVAL-g-PLGA. Pharm Res 20:2011–2020

14. Beck-Broichsitter M, Kleimann P, Gessler T et al (2012) Nebulization performance of biodegradable sildenafil-loaded nanoparticles using the Aeroneb® Pro: formulation aspects and nanoparticle stability to nebulization. Int J Pharm 422:398–408

15. Beck-Broichsitter M, Kleimann P, Schmehl T et al (2012) Impact of lyoprotectants for the stabilization of biodegradable nanoparticles on the performance of air-jet, ultrasonic, and vibrating-mesh nebulizers. Eur J Pharm Biopharm 82:272–280

16. Beck-Broichsitter M, Knuedeler MC, Schmehl T et al (2013) Following the concentration of polymeric nanoparticles during nebulization. Pharm Res 30:16–24

17. Dailey LA, Jekel N, Fink L et al (2006) Investigation of the proinflammatory potential of biodegradable nanoparticle drug delivery systems in the lung. Toxicol Appl Pharmacol 215:100–108

18. Beck-Broichsitter M, Schmehl T, Seeger W et al (2011) Evaluating the controlled release properties of inhaled nanoparticles using isolated, perfused, and ventilated lung models. J Nanomater. Article ID 163791.

19. Sakagami M (2006) In vivo, in vitro and ex vivo models to assess pulmonary absorption and disposition of inhaled therapeutics for systemic delivery. Adv Drug Deliv Rev 58:1030–1060

20. Agu RU, Ugwoke MI (2011) In vitro and in vivo testing methods for respiratory drug delivery. Expert Opin Drug Deliv 8:57–69

21. Nahar K, Gupta N, Gauvin R et al (2013) In vitro, in vivo and ex vivo models for studying particle deposition and drug absorption of inhaled pharmaceuticals. Eur J Pharm Sci 49:805–818

22. Vauthier C, Bouchemal K (2009) Methods for the preparation and manufacture of polymeric nanoparticles. Pharm Res 26:1025–1058

23. Beck-Broichsitter M, Rytting E, Lebhardt T et al (2010) Preparation of nanoparticles by solvent displacement for drug delivery: a shift in the "ouzo region" upon drug loading. Eur J Pharm Sci 41:244–253

24. Danhier F, Ansorena E, Silva JM et al (2012) PLGA-based nanoparticles: an overview of biomedical applications. J Contr Release 161:505–522

25. Dailey LA, Kissel T (2005) New poly(lactic-co-glycolic acid) derivatives: modular polymers with tailored properties. Drug Discov Today: Technol 2:7–14

26. Wittmar M, Unger F, Kissel T (2006) Biodegradable brushlike branched polyesters containing a charge-modified poly(vinyl alcohol) backbone as a platform for drug delivery systems: synthesis and characterization. Macromolecules 39:1417–1424

27. Hofmann W (2011) Modelling inhaled particle deposition in the human lung – a review. J Aerosol Sci 42:693–724

28. Beck-Broichsitter M, Schweiger C, Schmehl T et al (2012) Characterization of novel spray-dried polymeric particles for controlled pulmonary drug delivery. J Contr Release 158: 329–335

29. Packhäuser CB, Lahnstein K, Sitterberg J et al (2009) Stabilization of aerosolizable nano-carriers by freeze-drying. Pharm Res 26:129–138

30. Sitterberg J, Özcetin A, Ehrhardt C et al (2010) Utilising atomic force microscopy for the characterisation of nanoscale drug delivery systems. Eur J Pharm Biopharm 74:2–13

31. Clark AR (1995) The use of laser diffraction for the evaluation of aerosol clouds generated by medical nebulizers. Int J Pharm 115:69–78

32. Seeger W, Walmrath D, Grimminger F et al (1994) Adult respiratory distress syndrome: model systems using isolated perfused rabbit lungs. Methods Enzymol 233:549–584

Chapter 9

Antibody Labeling with Radioiodine and Radiometals

Suprit Gupta, Surinder Batra, and Maneesh Jain

Abstract

Antibodies have been conjugated to radionuclides for various in vitro and in vivo applications. Radiolabeled antibodies have been used in clinics and research for diagnostic applications both in vitro as reagents in bioassays and in vivo as imaging agents. Further, radiolabeled antibodies are used as direct therapeutic agents for cancer radioimmunotherapy or as tracers for studying the pharmacokinetics and biodistribution of therapeutic antibodies. Antibodies are labeled with radiohalogens or radiometals, and the choice of candidate radionuclides for a given application is dictated by their emission range and half-life. The conjugation chemistry for the coupling of MAbs with the radiometals requires a chelator, whereas radiohalogens can be incorporated directly in the antibody backbone. In this chapter, we describe the commonly used methods for radiolabeling and characterizing the antibodies most commonly used radiohalogens ($^{125}I/^{131}I$) and radiometals ($^{177}Lu/^{99m}Tc$).

Key words Antibodies, Radiometals, Radioiodine, Radiolabeling, Tracers, Iodogen, Chelator, DOTA, DTPA, 177Lu/99mTc

1 Introduction

Radiolabeled antibodies have diverse applications in biomedical research and clinical practice. Due to their ability to selectively target tumor antigens, radiolabeled monoclonal antibodies (MAbs) are used for the delivery of both diagnostic and therapeutic radionuclides in vivo for radioimmunodiagnosis and radioimmunotherapy, respectively. Further, several radiolabeled antibodies serve as critical reagents in radioimmunoassays for quantitative estimation of biomarkers in serum. The selection of radionuclide for antibody conjugation depends on the use of the radioimmunoconjugate and is dictated by the range of emission, emission type, and half-life of radionuclide [1]. Various radionuclides decay by emitting γ-radiation, β-particles, or α-particles. Due to their greater emission range, considerable penetration, and low linear energy transfer rates, β-emitters can kill surrounding cells by cross-fire effect and are thus used as therapeutic radionuclides [2]. ^{90}Y-a pure β-emitter and ^{131}I-a dual β and γ emitter are the

Kewal K. Jain (ed.), *Drug Delivery System*, Methods in Molecular Biology, vol. 1141,
DOI 10.1007/978-1-4939-0363-4_9, © Springer Science+Business Media New York 2014

only FDA-approved therapeutic radionuclides for conjugating antibodies for cancer therapy, while ^{111}In and [99m]Tc (γ emitters)-labeled MAbs have been approved for diagnostic applications. Due to its relatively long half-life and ease of handling, ^{125}I is also the radionuclide of choice for antibody-based radioimmunoassays, tracer studies for pharmacokinetics and biodistribution, and treatment of microscopic residual disease [3]. Due to its short half-life (6.7 day) and ability to emit both gamma and beta radiation, ^{177}Lu can be used simultaneously for therapy and diagnosis. Due to its shorter range of penetration than other β emitters, it has been explored for the treatment of smaller tumors in many clinical trials [4]. While predominantly intact IgGs are conjugated to radionuclides, various other formats including Fab' and scFvs have been used for various clinical and preclinical applications [5, 6].

The coupling of MAbs to a radionuclide depends upon the chemistry and half-life of radionuclide. Due to their easy availability, ease of handling, and relatively longer half-lives, radioisotopes of iodine (^{123}I, ^{125}I, ^{131}I) have been extensively used for labeling antibodies. The chemistry of iodine is well understood, and it can form stable covalent bonds causing minimal alteration to the protein backbone. Radioiodine is directly introduced by the halogenation (in the presence of enzymatic or chemical oxidants) of tyrosine and histidine residues of the MAbs [7]. Iodogen and chloramine-T are the most commonly used chemical oxidants employed for direct labeling and convert sodium iodide to iodine form, which spontaneously incorporates into tyrosyl groups of the proteins. In order to achieve higher labeling efficiency, the oxidant should be compatible with the aqueous solution of protein and should not affect the structure of the protein. In contrast to chloramine-T, iodogen method achieves lower specific activity but exhibits relatively milder effect on protein stability. Unlike iodination, conjugation of metallic radionuclides such as ^{90}Y, ^{111}In, ^{177}Lu, and ^{99m}Tc to antibodies requires a chelating agent. The selection of chelating agent largely depends on the physical properties and oxidation state of the radiometal ion to be conjugated. Usually, a bifunctional chelating agent (BFCA) is used which can bind covalently to MAbs on one hand and chelate radiometals on the other without affecting the kinetic and thermodynamic stability. The chelator provides the donor atoms which saturate the coordination sphere of the metal complex, thus stabilizing it. Several chelators like DOTA (1,4,7,10-tetraazacyclododecane-1,4,7,10-tetraacetic acid), DTPA (NR-diethylenetriaminepentaacetic acid), and NOTA (1,4,7-triazacyclononane-1,4,7-triacetic acid) have been used for radiolabeling antibodies for radioimmunotherapy and radioimmunodiagnosis. In this chapter, the labeling of antibody with heavy metal radionuclides (^{177}Lu, ^{99m}Tc) and radiohalogen (^{125}I) is described.

2 Materials

All solutions must be prepared in ultrapure water unless specified. A radiation safety manual should be consulted before handling any radioactive material (*see* **Note 1**).

2.1 Labeling with Radioiodine (See Note 2)

1. Iodogen (Pierce Chemical Co., Rockford).
2. Na^{125}I or Na^{131}I (New England Reactor, Boston, Massachusetts).
3. 10 mM sodium phosphate buffer: Add 3.1 g of NaH$_2$PO$_4$. H$_2$O and 10.9 g of Na$_2$HPO$_4$ to distilled water and make up the volume to 1 l. Set the pH of the solution to 7.2 and store at 4 °C.
4. 5 mM sodium Iodide: Dissolve 74.9 g of sodium iodide in 100 ml of ultrapure water and store at room temperature.
5. Chloroform.

2.2 Radiolabeling with ^{99m}Tc (See Note 3)

1. Tricine (Sigma-Aldrich): Dissolve 1 mg of tricine in 1 ml of ultrapure water to attain a concentration of 1 mg/ml and store at room temperature.
2. Stannous chloride dihydrate (Sigma-Aldrich): Dissolve 1 mg of stannous chloride in 1 ml of 0.1 N HCl to attain a concentration of 1 mg/ml and store at room temperature.
3. *N*-hydroxy succinimide sodium salt (NHS) (Pierce), stored dry at ambient temperature.
4. 20× PBS (phosphate buffered saline): Dissolve 160 g NaCl, 4 g KCl, 28.8 g NaH$_2$PO$_4$, and 4.8 g KH$_2$PO$_4$ in 600 ml of ultrapure water. Mix well, set pH to 7.4, and make up the volume to 1 l. For the working solution, add 50 ml to 950 ml of ultrapure water. This will give a working concentration of 137 mM NaCl, 2.7 mM KCl, 4.3 mM NaH$_2$PO$_4$, and 1.4 mM KH$_2$PO$_4$.
5. 10 mM sodium phosphate buffer: Add 3.1 g of NaH$_2$PO$_4$. H$_2$O and 10.9 g of Na$_2$HPO$_4$ to distilled water and make up the volume to 1 l. Set the pH of the solution to 7.8. The solution can be stored at 4 °C for up to 1 month.
6. 20 mM sodium citrate: Dissolve 5.88 g of sodium citrate dihydrate in 1 l of the water and store at room temperature.
7. 150 mM/l sodium acetate: Dissolve 12.30 g of anhydrous sodium acetate in 600 ml of ultrapure water. Set the pH of the solution to 7.8 and make up the volume to 1 l and store at room temperature.
8. 30 mM dimethylformamide.
9. ^{99m}Tc (supplied as pertechnetate-^{99m}TcO$_4$, fresh from ^{99m}Tc generator).

2.3 Radiolabeling with $^{Lu}177$ (See Note 3)

1. ITCB-DTPA (isothiocyanato-benzyl-diethylene pentaacetic acid) (Sigma, Poole, Dorset, UK): Prepare 5 mM aqueous solution.

2. ^{177}Lu (usually supplied as $^{177}Lu_2O_3$) (Oak Ridge National Laboratory, Oak Ridge, TN).

3. Chelex-100 Resin (Bio-Rad Laboratories, CA).

4. 20× PBS (phosphate buffered saline): Dissolve 160 g NaCl, 4 g KCl, 28.8 g NaH_2PO_4, and 4.8 g KH_2PO_4 in 600 ml of ultrapure water. Mix well, set pH to 7.4, and make up the volume to 1 l. For the working solution, add 50–950 ml of ultrapure water. This will give a working concentration of 137 mM NaCl, 2.7 mM KCl, 4.3 mM NaH_2PO_4, and 1.4 mM KH_2PO_4.

5. 0.05 M sodium carbonate: Dissolve 5.29 g of sodium carbonate in 600 ml of ultrapure water. Set the pH of the solution to 8.3, adjust the volume to 1 l, and store at room temperature.

6. 0.06 M sodium citrate: Dissolve 17.64 g of sodium citrate dihydrate in 600 ml of ultrapure water. Set the pH of the solution to 5.5 with 1 N HCl, adjust the volume to 1 l, and store at room temperature.

7. 0.6 M sodium acetate: Dissolve 49.21 g of sodium citrate dihydrate in 600 ml of ultrapure water. Set the pH of the solution to 5.3 with 1 N HCl and adjust the volume to 1 l. The solution can be stored at room temperature.

2.4 SDS-Polyacrylamide Gel Components

1. Resolving gel buffer (4× Tris–HCl pH 8.8): Dissolve 182 g of Tris base in 600 ml of water. Adjust pH to 8.8 with 1 N HCl and add water to make 1,000 ml. Filter the solution through 0.45 μm filter, add 2 g of SDS (sodium dodecyl sulfate), and store at 4 °C.

2. Stacking gel buffer (4× Tris–HCl pH 6.8): Dissolve 60.5 g of Tris base in 600 ml of water. Adjust pH to 6.8 with 1 N HCl and add water to make 1,000 ml. Filter the solution through 0.45 μm filter, add 4 g of SDS, and store at 4 °C.

3. 6× SDS sample buffer: To 7 ml of Tris–HCl pH 6.8, add 3 ml of glycerol, 1 g of SDS, and 0.5 ml of beta mercaptoethanol. Add 12 mg of bromophenol blue and mix it well. Make up the volume to 10 ml by water and store in –20 °C.

4. 30 % acrylamide solution (National Diagnostics).

5. 10 % ammonium persulfate (APS): Dissolve 100 mg of APS in 0.7 ml of water and adjust the volume to 1 ml. Prepare fresh for each use.

6. N,N,N′,N′-tetramethylethylenediamine (TEMED) (Fisher BioReagents).

7. SDS-running buffer: Add 12 g of Tris, 57.6 g of glycine, and 40 ml of 10 % SDS in 2.5 l of water and mix it well. Adjust volume to 4 l with water.

2.5 Coomassie Staining Components

1. *Staining solution*: Add 100 ml of glacial acetic acid to 500 ml of water. With constant stirring, add 400 ml of methanol and 1 g of Coomassie R250 dye and mix well. Filter with 0.45 μm filter and store at room temperature.

2. *Destaining solution*: Add 200 ml of methanol and 100 ml of glacial acetic acid in 700 ml of water and store at room temperature.

2.6 Instant Thin Layer Chromatography (ITLC) Components

1. ITLC-SG strips (silica impregnated glass fiber sheets).
2. Chromatography chamber.
3. Methanol.
4. 150 mM sodium acetate.

2.7 Other Components

1. Fume hood (SEFA 1-2010).
2. Gamma counter.
3. Dose calibrator (Capintec Inc., Ramsey, NJ).
4. Lead shielding.
5. Gel dryer.
6. Sephadex G-10 column and G-25 column (Pharmacia).
7. Microseparation filter (Centricon 30).
8. pH meter.
9. Sterile 12×75 mm glass tubes.
10. Glass beaker.
11. Centrifuge.
12. Eppendorf tubes.
13. Glass plates.
14. Whatman filter paper 3.
15. Kodak Film (Rochester, NY).
16. Light plus intensifying screen (Wilmington, DE).
17. X-Ray cassette.

3 Methods

3.1 Labeling of Antibody with ^{125}I [8]

1. Dissolve iodogen in chloroform to attain a concentration of 10 mg/ml.
2. Dispense 200 μl of iodogen solution in glass tubes and dry chloroform under a gentle stream of air while constantly swirling the tube to ensure uniform coating.
3. Cap the tube and store in –20 °C till further use. Iodogen-coated tubes can be stored for up to 1 year.

4. Equilibrate Sephadex G-25 10 ml column with 10 column volume of 0.1 M sodium phosphate buffer (or any desired buffer for downstream application of the antibody).

5. Place the iodogen-coated tube in the fume hood and allow it to come to room temperature. Add 10 μl of 100 mM sodium phosphate buffer (pH 8.0).

6. Adjust the concentration of antibody solution to 1 mg/ml and add 50–200 μl of the antibody to iodogen-coated tube containing sodium phosphate.

7. Behind an appropriate lead shielding in a fume hood, carefully open the vial containing radioiodine and determine the radioactivity/μl using a dose calibrator. Add 50–200 μCi of radioiodine (^{125}I or ^{131}I) to the bottom of the tube and gently swirl the tube. Typically 1 μCi radioiodine is added per μg of protein. However, if higher specific activity is desired, the ratio can be adjusted by adding more radioiodine (see **Note 4**).

8. Measure the total radioactivity in the reaction tube using dose calibrator. After 2–3 min incubation at room temperature, load the samples on the buffer-equilibrated Sephadex column to separate the iodinated antibody from the free iodine. Rinse the tube with 50–100 μl of sodium phosphate buffer and add the resulting solution to buffer.

9. Once the entire antibody-radioiodine reaction mixture has entered into the column matrix, add sodium phosphate buffer to fill the column reservoir and collect twenty 500 μl fractions in 5 ml (75×12 mm) plastic tubes. Measure the radioactivity in each fraction. The first peak represents radioiodinated protein, while the subsequent flat peak represents free iodine.

10. Cap the column and measure the residual activity using dose calibrator.

11. Pool the fractions of the iodinated antibody and store the labeled antibody at 4 °C.

12. Determine the efficiency of labeling from radioactivity measurements from **steps 8–11** and perform ITLC to determine free radioiodine.

13. Calculate the specific activity of the radiolabeled antibody (see **Note 5**).

3.2 Labeling of Antibody with ^{99m}Tc [9, 10]

3.2.1 Preparation of Antibody–Chelator Conjugate

1. Dissolve succinimidyl-6-hydrazinonicotinate hydrochloride (SHNH) in 30 mM dimethylformamide to prepare the hydrazinonicotinamide chelator at a concentration of 2–4 mg in 100–200 μl.

2. Dissolve 5 mg of IgG in 1 ml of 0.1 M sodium phosphate buffer pH 7.8.

3. With constant stirring, add 10 parts of modified SHNH to 1 part of IgG in 0.1 M sodium phosphate at 4 °C in dark.

4. Allow the reaction to occur overnight.

5. Set up the Sephadex G-10 column and equilibrate with 10 column volumes of 0.1 M sodium phosphate.

6. Purify the modified or bound protein from the unreacted fraction by loading the protein on column using 100 mM NaCl pH 5.2 buffered with 20 mM sodium citrate.

7. Collect the fractions as flow through in a fresh tube and pass through the column again.

8. Pool the fractions containing conjugated protein and concentrate the pooled fractions to 1 mg/ml using Centricon 100 centrifugal filters.

9. Store the SHNH–antibody conjugate at 4 °C till further use.

3.2.2 Radiolabeling of the Antibody

1. Aliquot 100 µg (100 µl) of tricine and 25 µg (25 µl) stannous chloride to fresh reaction tubes.

2. Using gamma counter, measure 1 mCi of ^{99m}Tc (sodium pertechnetate) and add to reaction tube described in **step 1**.

3. Allow the reaction to occur for 15 min at room temperature.

4. Add 400 µg of SHNH-derivatized IgG to the reaction tube containing ^{99m}Tc tricine and stannous chloride.

5. Allow the reaction to occur for 45 min at room temperature.

6. Set up the Sephadex G-25 column and equilibrate with 10 column volume of 0.1 M sodium phosphate.

7. Load the sample in column to separate the radiolabeled IgG from free ^{99m}Tc.

8. Elute the column with buffer consisting of 100 mM NaCl pH 7.5 buffered with 20 mM sodium citrate and collect fractions as described in **step 9** in Subheading 3.1.

9. Pool fraction corresponding to the radiolabeled protein and concentrate the pooled fractions to 1 mg/ml using a Centricon 100 by centrifugation.

10. Determine the labeling efficiency using ITLC.

3.3 Labeling of Antibody with ^{177}Lu [4]

3.3.1 Preparation of Antibody–Chelator Conjugate

1. Prepare the antibody in sodium carbonate buffer, pH 8.3, such that the final concentration of 5 mg/ml is achieved.

2. Add 33 µl aqueous solution of ITCB-DTPA to the above tube.

3. Allow the reaction to proceed for 2 h at room temperature.

4. Equilibrate Sephadex G-25 column with 10 column volume of 0.05 M sodium carbonate.

5. Separate the ITCB-DTPA-bound antibody fractions from the unreacted fractions by passing through the column.

6. Collect the fractions as flow through in a fresh tube and pass through the column again.

7. Pool the bound fractions by eluting with 100 mM PBS buffered with 20 mM sodium carbonate.

8. Adjust the immunoconjugate concentration to 10 mg/ml in PBS.

9. Aliquot the fractions into fresh tubes and store in –20 °C till further use.

3.3.2 Radiolabeling of the Antibody

1. Thaw 1 mg of the immunoconjugate and allow it to reach room temperature.

2. Transfer the content to fresh reaction tube.

3. Add 50 μl each of 0.6 M sodium acetate and 0.06 M sodium citrate to the above tube.

4. Measure 1 mCi activity of ^{177}Lu using a dose calibrator and add to the tube using metal-free pipette tips.

5. Allow the reaction to occur for 2 h at room temperature.

6. Equilibrate Sephadex G-25 column with 10 column volume of 0.05 M sodium carbonate and load the sample to separate the ^{177}Lu-bound hot fractions from the unbound one.

7. Collect the fractions as flow through in a fresh tube and pass through the column again as described in **step 9** Subheading 3.1.

8. Pool the bound fractions by eluting with 100 mM PBS buffered with 20 mM sodium carbonate.

9. Concentrate the pooled fractions to 1 mg/ml concentration using Centricon 100.

10. Check the labeling efficiency using ITLC.

11. Other chelators can also be used (*see* **Note 6**).

3.4 Assessment of Radiochemical Purity Using ITLC

1. Cut ITLC-SC sheet into narrow strips (1×10 cm).

2. On each side of ITLC strips, mark with a soft pencil an origin (approximately 1 cm from the bottom of the strip).

3. Using a water-soluble marker, place a small dot 1 cm below the upper edge of the ITLC strip (this helps to follow the progress of the elution: remove the strip from the developing chamber when the ink begins to run).

4. Place 1–2 μl of column purified radiolabeled antibody (from pooled fractions) at the origin of the ITLC strip and allow it to air-dry. Run triplicate ITLC strips radioimmunoconjugate.

5. Place the strips carefully in chromatography chamber containing appropriate solvent (meniscus not higher than 0.5 cm from the bottom) such that the bottom touches the solvent and strip lies on the chamber wall. Cover the chamber with the lid (*see* **Note 7**).

6. Allow the solvent to reach the ink dot, remove strips from the developing chamber, and allow to air-dry (approximately 2 min).

7. Cut the ITLC strip into two equal top and bottom parts. Bottom contains origin with protein-bound radioactivity, while top contains solvent front with free radionuclide.

8. Place the top and bottom parts in in two separate tubes and measure radioactivity using gamma counter.

9. Calculate percent protein-bound radioactivity according to the formula listed below (*see* **Note 8**):

$$\frac{CPM^{bottom} \times 100}{\left(CPM^{top} + CPM^{bottom}\right)}.$$

3.5 Gel Electrophoresis

1. Perform an SDS-polyacrylamide gel electrophoresis (PAGE) under reducing and nonreducing conditions [11].

2. Following electrophoresis, remove the gel from glass plate and rinse with ultrapure water.

3. Add Coomassie staining solution and put for 1 h at room temperature under mild shaking conditions.

4. Add destaining solution, replacing the solution by every 15–20 min until faint bands are seen. Continue destaining the gel till bands are clean.

5. Rinse the gel with ultrapure water once.

6. With the help of Whatman filter paper, carefully remove the gel and place on a gel dryer.

7. Allow the gel to dry for 2 h at 80 °C.

8. Place the gel in an X-ray cassette and expose the gel to an autoradiography film overnight.

9. Develop the film to visualize protein bands. A single band indicating intact antibody should be visible under nonreducing conditions, while two bands corresponding to antibody heavy and light chains should be visible under reducing conditions. There should be minimal signal near the dye front (indicating free radionuclide) (*see* **Note 9**).

4 Notes

1. Before using radioactive isotopes, consult the radiation safety office for proper handling, usage, and disposable of radionuclides.

2. Free radioiodine (NaI) is volatile and should only be handled in a fume hood. In general, all labeling reactions must be performed in fume hood with appropriate lead shielding.

3. For labeling with radiometals, all reagents should be prepared in prepared in metal-free water using metal-free glassware and pipettes. Metal ions from water and reagents can be eliminated either by passing them through Chelex-100 column or by addition resin directly to the reagents.

4. If dose calibrator is not available, radioactivity can be measured using gamma counter, and CPM can be converted to Ci, mCi, or μCi. First, convert CPM (counts per minute) to DPM (disintegrations per minute) as follows:

$$DPM = \frac{CPM_{sample} - CPM_{background}}{Detector\ Efficiency}.$$

The background CPM and detector efficiency should be determined for gamma counter as per manufacturer's instructions:

$$1\ \mu\ Ci = 2.22 \times 10^6\ DPM.$$

5. Specific activity is the amount of radioactivity per unit mass of protein. To determine specific activity, the amount of radioactivity in the radiolabeled protein must be measured using a gamma counter, and the protein concentration should be determined using any standard protein estimation method (BCA, Bradford).

6. Other chelators can also be used for radiometal labeling of antibodies, as detailed in other publications [12, 13].

7. Methanol/water (1:4 v/v) is used as a solvent for ITLC of radioiodinated antibody, and 0.15 mM sodium acetate is used for ^{99m}Tc. For ^{177}Lu, parallel ITLC strips should be run in methanol/water and sodium acetate.

8. The amount of free radionuclide should not exceed more than 5 %. Excess free label should be removed using Sephadex 25 column.

9. Immunoreactivity of the radiolabeled antibody should be ascertained using appropriate immunoassay established in the laboratory (solid phase RIA, ELISA, or immunoblotting).

Acknowledgements

The authors on this work are supported, in part, by grants from the National Institutes of Health P20GM103480, R21 CA156037, U01CA111294, R03 CA 139285, R03 CA167342, P50 CA127297, and U54163120.

References

1. Milenic DE, Brady ED, Brechbiel MW (2004) Antibody-targeted radiation cancer therapy. Nat Rev Drug Discov 3:488–499

2. O'Donoghue JA, Bardies M, Wheldon TE (1995) Relationships between tumor size and curability for uniformly targeted therapy with beta-emitting radionuclides. J Nucl Med 36: 1902–1909

3. Makrigiorgos G, Adelstein SJ, Kassis AI (1990) Auger electron emitters: insights gained from in vitro experiments. Radiat Environ Biophys 29:75–91

4. Chauhan SC, Jain M, Moore ED, Wittel UA, Li J, Gwilt PR, Colcher D, Batra SK (2005) Pharmacokinetics and biodistribution of 177Lu-labeled multivalent single-chain Fv construct of the pancarcinoma monoclonal antibody CC49. Eur J Nucl Med Mol Imaging 32:264–273

5. Batra SK, Jain M, Wittel UA, Chauhan SC, Colcher D (2002) Pharmacokinetics and biodistribution of genetically engineered antibodies. Curr Opin Biotechnol 13:603–608

6. Jain M, Kamal N, Batra SK (2007) Engineering antibodies for clinical applications. Trends Biotechnol 25:307–316

7. Anderson WT, Strand M (1987) Radiolabeled antibody: iodine versus radiometal chelates. NCI Monogr 3:149–151

8. Colcher D, Zalutsky M, Kaplan W, Kufe D, Austin F, Schlom J (1983) Radiolocalization of human mammary tumors in athymic mice by a monoclonal antibody. Cancer Res 43: 736–742

9. Abrams MJ, Juweid M, TenKate CI, Schwartz DA, Hauser MM, Gaul FE, Fuccello AJ, Rubin RH, Strauss HW, Fischman AJ (1990) Technetium-99m-human polyclonal IgG radiolabeled via the hydrazino nicotinamide derivative for imaging focal sites of infection in rats. J Nucl Med 31:2022–2028

10. Goel A, Baranowska-Kortylewicz J, Hinrichs SH, Wisecarver J, Pavlinkova G, Augustine S, Colcher D, Booth BJ, Batra SK (2001) 99mTc-labeled divalent and tetravalent CC49 single-chain Fv's: novel imaging agents for rapid in vivo localization of human colon carcinoma. J Nucl Med 42:1519–1527

11. Laemmli UK (1970) Cleavage of structural proteins during the assembly of the head of bacteriophage T4. Nature 227:680–685

12. Hens M, Vaidyanathan G, Welsh P, Zalutsky MR (2009) Labeling internalizing anti-epidermal growth factor receptor variant III monoclonal antibody with (177)Lu: in vitro comparison of acyclic and macrocyclic ligands. Nucl Med Biol 36:117–128

13. Hens M, Vaidyanathan G, Zhao XG, Bigner DD, Zalutsky MR (2010) Anti-EGFRvIII monoclonal antibody armed with 177Lu: in vivo comparison of macrocyclic and acyclic ligands. Nucl Med Biol 37:741–750

Chapter 10

Self-Assembling Peptide-Based Delivery of Therapeutics for Myocardial Infarction

Archana V. Boopathy and Michael E. Davis

Abstract

Drug and cell delivery systems could be modulated to serve as instructive microenvironments in regenerative medicine. Towards this end, several synthetic biomaterials have been developed to mimic the natural extracellular matrix (ECM) for therapeutic use. These include synthetic polymers, decellularized ECM, self-assembling polymers, and cell-responsive hydrogels with varied applications. Here, we describe the development of a self-assembling peptide hydrogel and its potential use as a cell and growth factor delivery vehicle to the infarcted heart in a rodent model of myocardial infarction.

Key words Self-assembling peptide, Hydrogel, Myocardial infarction, Cell therapy

1 Introduction

Self-assembling peptides have widespread applications in regenerative medicine, cell therapy, and drug delivery. Self-assembling peptides are composed of alternating hydrophilic (arginine or aspartic acid) and hydrophobic (alanine or lysine) amino acids with the peptide sequence H_2N-RARADADARARADADA-OH. Due to the charges on the hydrophilic groups when the peptides are aligned, a stable antiparallel β-sheet is formed that at neutral pH polymerizes into a hydrogel [1]. The peptides self-assemble into peptide nanofibers (7–20 nm in diameter) with >99 % water content at physiological pH and osmolarity [2]. The peptide hydrogel has been shown to be noninflammatory and non-immunogenic. The peptide sequence does not contain any cell recognizable structures and is considered to be a noninstructive hydrogel. Further, the 3D scaffold environment also promotes attachment, proliferation, and differentiation of different cell types: neuronal cells [3], endothelial cells [4], hepatocytes [5], chondrocytes [6], osteoblasts [7], and progenitor cells [8, 9]. During solid-phase synthesis, the peptides can be functionalized with ligands and adhesive motifs at the C-terminal to provide signaling cues and promote cell retention in

Kewal K. Jain (ed.), *Drug Delivery System*, Methods in Molecular Biology, vol. 1141,
DOI 10.1007/978-1-4939-0363-4_10, © Springer Science+Business Media New York 2014

the scaffold [2, 10]. The self-assembling peptides can be functionalized with (1) short motifs for cell adhesion to modulate human adipose stem cell behavior [11], (2) VEGF to promote cardiac repair following infarction in pigs [12], and (3) growth factors through biotin–streptavidin linkages for use in cell therapy [10]. The hydrogels have also been used for cartilage repair, axon regeneration, treatment of bone defects, and myocardial infarction [8, 13]. After synthesis, the peptides can be purified by high-pressure liquid chromatography on reverse-phase columns and the presence of a correct composition verified by amino acid analysis.

2 Materials

2.1 Self-Assembling Hydrogel

1. Self-assembling peptide with the amino acid sequence H_2N-RARADADARARADADA-OH. The peptide is synthesized as crude ammonium acetate salt. Purity and composition is verified by reverse-phase HPLC and mass spectroscopy.
2. Sucrose—Prepare a 295 mM solution in water. Sterile filter through a 0.22 μm filter. Store at 4 °C until use.
3. Sonicator (Fisher Scientific Sonic Dismembrator Model 100).
4. Weigh balance with sensitivity to measure a minimum of 0.5 mg (Denver Instrument APX-60).

2.2 Cell Embedding

1. Cell line or primary cells in culture.
2. 0.25 % Trypsin–EDTA (Gibco).
3. Culture media.
4. Transwell inserts (Cat# PICM01250 Millipore; 0.4 μm pore size, 12 mm diameter).

2.3 In Vivo Injection

1. Rodent approved for the study.
2. Appropriate surgical instruments.
3. Electric clipper with #40 blade.
4. Isoflurane solution.
5. 6–0 monofilament suture.
6. Prolene suture.
7. ½ cc 30 G Ultra-fine insulin syringe (BD Cat# 328468).

2.4 Instruments and Plasticware

All plasticware must be sterile.

1. 24-Multiwell plates (Corning).
2. Cell culture incubator at 37 °C, 5 % CO_2 (Thermo Scientific).
3. Laminar flow hood (Thermo Scientific).
4. Light microscope (Nikon TMS).

5. Plastic pipettes (Costar Incorporated).

6. 15 ml plastic conical tubes (Falcon).

7. Pipettes: 10, 200 and 1,000 μl (Denville).

8. Pipette tips: 10, 200 and 1,000 μl (Denville).

9. Benchtop centrifuge (Thermo Scientific).

10. 37 °C water bath (Thermo Scientific).

3 Methods

Carry out all procedures at room temperature.

3.1 Preparation of the Hydrogel

1. Warm sucrose solution to room temperature.

2. Obtain the self-assembling peptide from a peptide synthesizing facility or a commercial source (*see* **Note 1**).

3. Weigh 1 mg of the self-assembling peptide to prepare a 1 % w/v solution in a 1.5 ml microcentrifuge tube (*see* **Notes 3** and **4**). Prepare hydrogels of desired percentage (1–3 % w/v) by adjusting the corresponding volume of sucrose solution added.

4. Centrifuge the tubes containing the peptide for 5 s to pellet the peptide.

5. Add 100 μl of sucrose solution to the tube containing 1 mg of peptide.

6. Sonicate for 30 s till the peptide dissolves to form a clear liquid (*see* **Note 2**).

7. Pipette the peptide solution into a Transwell insert in a 24-well plate.

8. Incubate at 37 °C for 15 min for the solution to polymerize into a hydrogel. The time required for polymerization and the mechanical properties of the hydrogel depend on the initial % concentration of the peptide used.

3.2 3D Culture of Cells Within the Hydrogel

Carry out all procedures under a sterile laminar flow hood.

1. Trypsinize the required cells from the culture flask (*see* **Note 5**).

2. Centrifuge the trypsin containing cell solution at $750 \times g$ for 5 min.

3. Count the number of cells using a hemocytometer (*see* **Note 6**).

4. Aliquot the required number of cells (usually 3–400,000) to be embedded per gel into a separate microcentrifuge tube.

5. Centrifuge the tubes to obtain a cell pellet.

6. Add the 100 μl of peptide solution (from **step 6** in Subheading 3.1) to the cell pellet.

7. Mix gently with a pipette without creating air bubbles (*see* **Note 7**).

8. For in vitro studies:

   (a) Transfer the cell containing peptide solution into a Transwell insert on a 24-well plate.

   (b) Incubate at 37 °C for 15 min for the solution to polymerize into a hydrogel. Add 200 μl of cell culture media inside and 300 μl media on the outside of the insert in the well for culturing the cells in the hydrogel (*see* **Note 8**).

   (c) Culture cells in the hydrogel for further experimentation.

9. For in vivo studies:

   (a) Transfer the cell containing peptide solution into a ½ cc 30 G insulin syringe immediately before injection.

   (b) For myocardial delivery, follow Subheading 3.3.

3.3 Myocardial Delivery of the Hydrogel

Obtain necessary institutional approvals and modify the protocol below accordingly.

1. Prepare adult Sprague–Dawley rats for surgery.

2. Anesthetize the rats using 5 % isoflurane initially and reduce to 1.5 % for maintenance during surgery.

3. Remove hair from the chest using electric clippers with a #40 blade.

4. Clean the chest area with a disinfectant solution (Betadine) followed by alcohol as the final scrub.

5. Place the animal on the heating pad and cover with drape except the site of surgery (*see* **Note 9**).

6. Insert the endotracheal tube and PE180 and mechanically ventilate at 110–120 RPM with 1–2 l/min of oxygen supply.

7. Expose the heart by separation of the ribs (*see* **Note 10**).

8. Induce myocardial infarction by ligating the left ascending coronary artery using a 6–0 monofilament suture.

9. After 30 min, remove the suture on the ligated artery to allow for reperfusion.

10. Inject the self-assembling peptide solution (maximum volume 50 μl) into the free wall of the left ventricle (LV) through a 30-G needle in three equal parts around the border zone of the infarction.

11. Once the injections are complete, close the chest using a Prolene suture and allow the animal to recover on a heating pad (*see* **Note 11**).

12. Administer analgesic as approved by the institutional animal care committee.

13. Monitor recovery of the animal over time.

4 Notes

1. The purity of the final peptide product must be >95 %.

2. Sonicate only 2 mg per tube for obtaining a maximum homogenous solution.

3. Weigh peptides directly in 1.5 ml microcentrifuge tubes. Calculate the difference in weights before and after peptide addition to determine the amount of peptide in the tube.

4. It is recommended to weigh excess of the peptide as not all of the peptide solution can be transferred out of the tube.

5. During trypsinization of cells, avoid foaming and bubbles.

6. Viability after trypsinization should be at least 80 % of the total cell number.

7. Mix the cells with the self-assembling peptide gently but quickly (1) to avoid self-assembling in the microcentrifuge tube, (2) to prevent cell death from repeated pipetting, and (3) to transfer all the cell containing peptide solution to the cell culture insert (for in vitro) or 30 G syringe (for in vivo).

8. Use serum-free or low-serum cell culture media while culturing cells in the hydrogel to avoid adsorption of serum proteins to the hydrogel.

9. Maintain the surgical area clean and uncluttered to minimize risk of infection following injection.

10. Clean and autoclave all surgical instruments before use.

11. Monitor animals post operation every 15 min for the first hour and every hour during the day for potential signs of pain or distress.

Acknowledgements

This work was supported by grant HL094527 to M.E.D. from the National Heart, Lung, and Blood Institute, an American Heart Association predoctoral fellowship 11PRE7840078 to A.V.B.

References

1. Hammond NA, Kamm RD (2013) Mechanical characterization of self-assembling peptide hydrogels by microindentation. J Biomed Mater Res B Appl Biomater 101:981–990

2. Segers VF, Lee RT (2007) Local delivery of proteins and the use of self-assembling peptides. Drug Discov Today 12:561–568

3. Semino CE, Kasahara J, Hayashi Y, Zhang S (2004) Entrapment of migrating hippocampal neural cells in three-dimensional peptide nanofiber scaffold. Tissue Eng 10:643–655

4. Genove E, Shen C, Zhang S, Semino CE (2005) The effect of functionalized self-assembling peptide scaffolds on human aortic endothelial cell function. Biomaterials 26:3341–3351

5. Genove E, Schmitmeier S, Sala A, Borros S, Bader A, Griffith LG, Semino CE (2009)

Functionalized self-assembling peptide hydrogel enhance maintenance of hepatocyte activity in vitro. J Cell Mol Med 13:3387–3397

6. Kisiday J, Jin M, Kurz B, Hung H, Semino C, Zhang S, Grodzinsky AJ (2002) Self-assembling peptide hydrogel fosters chondrocyte extracellular matrix production and cell division: implications for cartilage tissue repair. Proc Natl Acad Sci U S A 99:9996–10001

7. Bokhari MA, Akay G, Zhang S, Birch MA (2005) The enhancement of osteoblast growth and differentiation in vitro on a peptide hydrogel-polyHIPE polymer hybrid material. Biomaterials 26:5198–5208

8. Padin-Iruegas ME, Misao Y, Davis ME, Segers VF, Esposito G, Tokunou T, Urbanek K, Hosoda T, Rota M, Anversa P, Leri A, Lee RT, Kajstura J (2009) Cardiac progenitor cells and biotinylated insulin-like growth factor-1 nanofibers improve endogenous and exogenous myocardial regeneration after infarction. Circulation 120:876–887

9. Pendergrass KD, Boopathy AV, Seshadri G, Maiellaro-Rafferty K, Che PL, Brown ME, Davis ME (2013) Acute preconditioning of cardiac progenitor cells with hydrogen peroxide enhances angiogenic pathways following ischemia-reperfusion injury. Stem Cells Dev 22(17):2414–2424

10. Davis ME, Hsieh PC, Takahashi T, Song Q, Zhang S, Kamm RD, Grodzinsky AJ, Anversa P, Lee RT (2006) Local myocardial insulin-like growth factor 1 (IGF-1) delivery with biotinylated peptide nanofibers improves cell therapy for myocardial infarction. Proc Natl Acad Sci U S A 103:8155–8160

11. Liu X, Wang X, Wang X, Ren H, He J, Qiao L, Cui FZ (2013) Functionalized self-assembling peptide nanofiber hydrogels mimic stem cell niche to control human adipose stem cell behavior in vitro. Acta Biomater 9:6798–6805

12. Lin YD, Luo CY, Hu YN, Yeh ML, Hsueh YC, Chang MY, Tsai DC, Wang JN, Tang MJ, Wei EI, Springer ML, Hsieh PC (2012) Instructive nanofiber scaffolds with VEGF create a microenvironment for arteriogenesis and cardiac repair. Sci Transl Med 4(146):146ra109

13. Kyle S, Aggeli A, Ingham E, McPherson MJ (2009) Production of self-assembling biomaterials for tissue engineering. Trends Biotechnol 27:423–433

Chapter 11

Applications of Chitosan Nanoparticles in Drug Delivery

H.A. Tajmir-Riahi, Sh. Nafisi, S. Sanyakamdhorn, D. Agudelo, and P. Chanphai

Abstract

We have reviewed the binding affinities of several antitumor drugs doxorubicin (Dox), N-(trifluoroacetyl) doxorubicin (FDox), tamoxifen (Tam), 4-hydroxytamoxifen (4-Hydroxytam), and endoxifen (Endox) with chitosan nanoparticles of different sizes (chitosan-15, chitosan-100, and chitosan-200 KD) in order to evaluate the efficacy of chitosan nanocarriers in drug delivery systems. Spectroscopic and molecular modeling studies showed the binding sites and the stability of drug–polymer complexes. Drug–chitosan complexation occurred via hydrophobic and hydrophilic contacts as well as H-bonding network. Chitosan-100 KD was the more effective drug carrier than the chitosan-15 and chitosan-200 KD.

Key words Nanoparticles, Chitosan, Drug delivery, Polymer, Binding site, Stability, Spectroscopy, Modeling

Abbreviations

Ch	Chitosan
Dox	Doxorubicin
FDox	N-(trifluoroacetyl) doxorubicin
Tam	Tamoxifen
4-Hydroxytam	4-Hydroxytamoxifen
Endox	Endoxifen
PEG	Poly(ethylene glycol)
FTIR	Fourier transform infrared

1 Introduction

Biodegradable and biocompatible chitosan (Scheme 1) and its derivatives have received major attention for the delivery of therapeutic drugs, proteins, and antigens [1–3]. Chitosan is a natural polymer obtained by a partial deacetylation of chitin [4]. It is nontoxic, biocompatible, and biodegradable polysaccharide.

Kewal K. Jain (ed.), *Drug Delivery System*, Methods in Molecular Biology, vol. 1141,
DOI 10.1007/978-1-4939-0363-4_11, © Springer Science+Business Media New York 2014

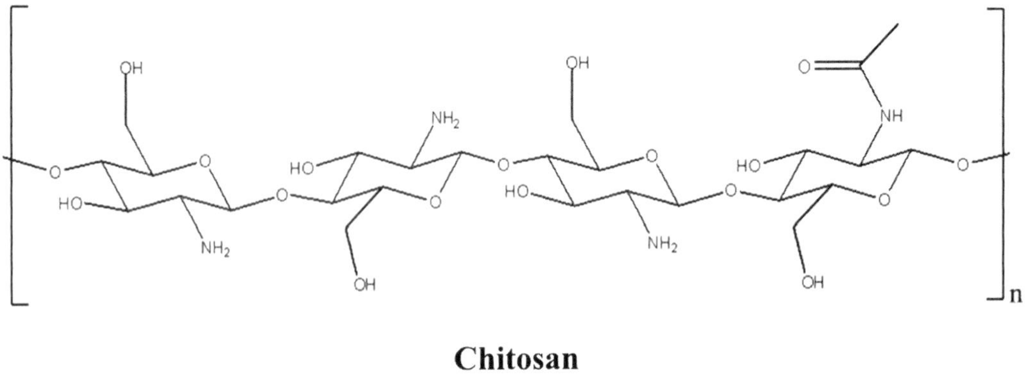

Chitosan

Scheme 1 Chemical structure of chitosan

Chitosan nanoparticles have gained more attention as drug delivery carriers because of their better stability, low toxicity, simple and mild preparation method, and providing versatile routes of administration [4–7]. The deacetylated chitosan backbone of glucosamine units has a high density of charged amine groups, permitting strong electrostatic interactions with proteins and genes that carry an overall negative charge at neutral pH conditions [4, 5]. The fast-expanding research of the useful physicochemical and biological properties of chitosan has led to the recognition of the cationic polysaccharide, as a natural polymer for drug delivery [8, 9]. The encapsulation of several drugs and proteins with chitosan nanoparticles is recently reported, and major hydrophilic and hydrophobic contacts as well as H-bonding were dominated in drug–polymer complexation [3, 10–13]. Therefore, a comparative study of drug encapsulation with chitosan of different sizes will be of a major interest in order to evaluate the efficacy of chitosan nanoparticles in drug delivery.

Among drugs used to evaluate the efficacy of chitosan nanoparticles as nanocarriers were doxorubicin and its analogue FDOX known as antitumor agents against several types of cancers such as acute leukemia, malignant lymphoma, and breast cancer [14–16]. Similarly, tamoxifen and its metabolites 4-hydroxytamoxifen and endoxifen known as breast cancer drugs were used in this evaluation [17–20].

We have reviewed here the important role of chitosan nanoparticles as drug delivery tools, using doxorubicin, N-(trifluoroacetyl) doxorubicin, tamoxifen, 4-hydroxytamoxifen, and endoxifen with chitosan of different sizes: chitosan-15, chitosan-100, and chitosan-200 KD. The analysis of the drug binding sites and the stability of drug–polymer complexes were compared here in order to evaluate the efficacy of chitosan nanoparticles in drug delivery systems.

2 Materials

1. Purified chitosans 15, 100, and 200 KD (90 % deacetylation) (Polysciences, Inc., Warrington, USA) and used as supplied.

2. Doxorubicin hydrochloride (Pharmacia/Farmitalia Carlo Erba, Italy) and N-(trifluoroacetyl) doxorubicin were synthesized according to the published methods [21, 22].

3. Tamoxifen and 4-hydroxytamoxifen (Sigma Chemical Company) were used as supplied.

4. Synthesis of endoxifen was conducted at the Chemical Synthesis Core Facility by Fauq et al. [23].

5. Since chitosan is not soluble in aqueous solution at neutral pH (soluble in 0.1 N acetic acid or HCl), an appropriate amount of chitosan was dissolved in acetate solution (pH 5.5–6.5). Drug solutions were prepared in ethanol/water (25/75 %) and diluted in acetate buffer.

3 Methods

3.1 FTIR Spectroscopic Measurements

Infrared spectra were recorded on an FTIR spectrometer (Impact 420 model, Digilab), equipped with deuterated triglycine sulfate (DTGS) detector and KBr beam splitter, using AgBr windows. The solution of drug was added dropwise to the chitosan solution with constant stirring to ensure the formation of homogeneous solution and to reach the target drug concentrations of 15, 30, and 60 μM with a final chitosan concentration of 30 μM. Spectra were collected after 2 h incubation of chitosan with drug solution at room temperature, using hydrated films. Interferograms were accumulated over the spectral range of 4,000–600 cm^{-1} with a nominal resolution of 2 cm^{-1} and 100 scans. The difference spectra [(chitosan solution + drug solution) – (chitosan solution)] were generated using free chitosan band around 902 cm^{-1}, as standard. This band is related to chitosan ring stretching [24, 25] and does not show alterations upon drug complexation. When producing difference spectra, this band was adjusted to the baseline level, in order to normalize the difference spectra.

3.2 Fluorescence Spectroscopy

Fluorimetric experiments were carried out on a PerkinElmer LS55 spectrometer. Stock solution of drug (30 μM) in acetate (pH 5.5–6.5) was also prepared at 24 ± 1 °C. Various solutions of chitosan (1–200 μM) were prepared from the above stock solutions by successive dilutions at 24 ± 1 °C. Samples containing 0.06 ml of the above drug solution and various polymer solutions were mixed to obtain final chitosan concentrations ranging from 1 to 200 μM with constant drug content (30 μM). The fluorescence spectra were recorded at $\lambda_{ex} = 480$ nm and λ_{em} from 500 to 750 nm. The intensity of the bands at 592 nm from doxorubicin and its analogue [26] and at 360 nm for tamoxifen and its metabolites [27] was used to calculate the binding constant (K) according to previous reports [28–33].

On the assumption that there are (n) substantive binding sites for quencher (Q) on protein (B), the quenching reaction can be shown as follows:

$$nQ + B \Leftrightarrow Q_n B. \tag{1}$$

The binding constant (K_A) can be calculated as

$$K_A = [Q_n B] / [Q]^n [B], \tag{2}$$

where $[Q]$ and $[B]$ are the quencher and protein concentration, respectively, $[Q_n B]$ is the concentration of nonfluorescent fluorophore–quencher complex, and $[B_0]$ gives the total protein concentration:

$$[Q_n B] = [B_0] - [B]. \tag{3}$$

$$K_A = ([B_0] - [B]) / [Q]^n [B]. \tag{4}$$

The fluorescence intensity is proportional to the protein concentration as described:

$$[B] / [B_0] \propto F / F_0. \tag{5}$$

Results from fluorescence measurements can be used to estimate the binding constant of drug–polymer complex. From Eq. 4,

$$\log[(F_0 - F) / F] = \log K_A + n \log[Q]. \tag{6}$$

The accessible fluorophore fraction (f) can be calculated by modified Stern–Volmer equation:

$$F_0 / (F_0 - F) = 1 / fK[Q] + 1 / f, \tag{7}$$

where F_0 is the initial fluorescence intensity and F is the fluorescence intensities in the presence of quenching agent (or interacting molecule). K is the Stern–Volmer quenching constant, $[Q]$ is the molar concentration of quencher, and f is the fraction of accessible fluorophore to a polar quencher, which indicates the fractional fluorescence contribution of the total emission for an interaction with a hydrophobic quencher [34, 35]. The K will be calculated from $F_0/F = K[Q] + 1$.

3.3 Molecular Modeling

The docking studies were carried out with ArgusLab 4.0.1 software (Mark A. Thompson, Planaria Software LLC, Seattle, WA, http://www.arguslab.com). The chitosan structure was obtained from a literature report [36], and the drug three-dimensional structures were generated from PM3 semiempirical calculations using Chem3D Ultra 11.0. The whole polymer was selected as a potential binding site since no prior knowledge of such site was available in the literature. The docking runs were performed on the ArgusDock docking engine using regular precision with a maximum of 150 candidate poses. The conformations were ranked using the Ascore scoring function, which estimates the free binding energy. Upon location of the potential binding sites, the docked complex conformations were optimized using a steepest decent algorithm until convergence, with a maximum of 20 iterations. Chitosan donor groups within a distance of 3.5 Å [37] relative to the drug were involved in complex formation.

4   Notes

1. Infrared difference spectroscopy was often used to characterize the nature of drug–polymer interactions [11, 12]. The major spectral shifting for the chitosan amide I band at 1,633–1,620 cm^{-1} (mainly C=O stretch) and amide II band at 1,540–1,520 cm^{-1} (C–N stretching coupled with N–H bending modes) [24, 25] together with the intensity changes of these bands obtained from difference spectra [(chitosan + drug solution) – (chitosan solution)] was used to analyze the nature of drug–chitosan bindings, and the results are shown in Figs. 1, 2, and 3. Similarly, the infrared spectral changes of the free chitosan in the region of 3,500–2,800 cm^{-1} were compared with those of the drug–polymer adducts in order to determine the drug binding to polymer OH and NH_2 groups, as well as the presence of hydrophobic and hydrophilic contacts in drug–chitosan complexes (Fig. 4).

   The major increase in the intensity of chitosan amide I at 1,633–1,620 cm^{-1} and amide II at 1,540–1,520 cm^{-1} in the difference spectra of the drug–polymer complexes was used as the marker for drug–polymer interaction via chitosan NH_2, N–H, and C=O groups (hydrophilic contacts). The positive features located at 1,660–1,620 cm^{-1} in the difference spectra of chitosan-15, chitosan-100, and chitosan-200 KD complexes with Dox, FDox, Tam, 4-Hydroxytam, and endoxifen are related to the increase in intensity of chitosan amide I band due to major drug–polymer hydrophilic interactions (Figs. 1, 2, and 3; diffs, 60 μM) [11, 12].

   The analysis of the infrared spectra of chitosan in the region of 3,500–2,800 cm^{-1} showed major shifting of polymer OH, N–H, and CH stretching modes (Fig. 4). The polymer OH stretching vibrations at 3,460, 3,427 (free ch-15), 3,423 (free ch-100), and 3,449, 3,390 cm^{-1} (free ch-200) showed major shifting and intensity changes in the spectra of

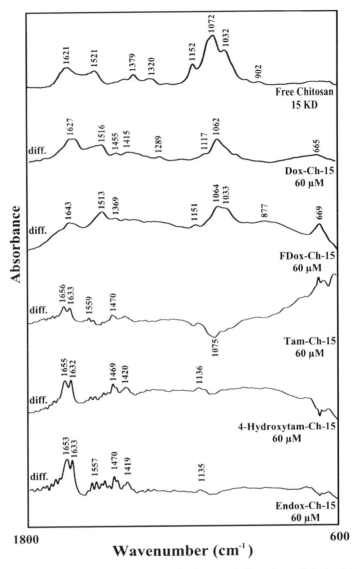

Fig. 1 FTIR spectra and difference spectra [(chitosan solution + drug solution) − (chitosan solution)] in the region of 1,800–600 cm⁻¹ for the free chitosan-15 and its drug complexes for Dox-ch-15, FDox-ch-15, tamoxifen-ch-15, 4-hydroxytamoxifen-ch-15, and endoxifen-ch-15 in aqueous solution at pH 5–6 with 60 µM drug and chitosan concentrations

Dox and FDox, Tam, 4-Hydroxytam, and endoxifen chitosan complexes (Fig. 4a–c). Similarly, the N–H stretching vibrations at 3,230 (free ch-15), 3,245 (free ch-100), and 3,239 cm⁻¹ (free ch-200) exhibit shifting upon drug complexation (Fig. 4a–c). The spectral changes of the polymer OH and N–H stretching modes are due to the participation of chitosan OH and NH₂ group in drug–polymer complexes (hydrophilic contacts) [11, 12].

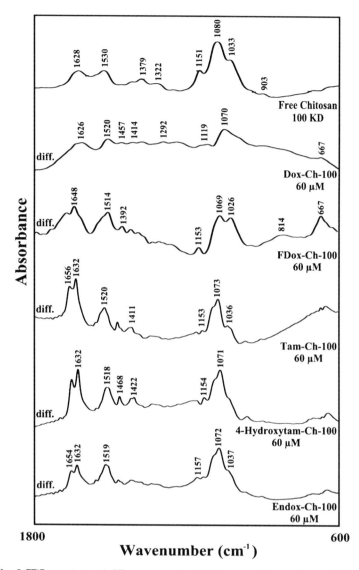

Fig. 2 FTIR spectra and difference spectra [(chitosan solution + drug solution) − (chitosan solution)] in the region of 1,800–600 cm⁻¹ for the free chitosan-100 and its drug complexes for Dox-ch-100, FDox-ch-100, tamoxifen-ch-100, 4-hydroxytamoxifen-ch-100, and endoxifen-ch-100 in aqueous solution at pH 5–6 with 60 μM drug and chitosan concentrations

Hydrophobic interactions were also characterized by the shifting of the chitosan symmetric and antisymmetric CH stretching vibrations observed at 2,979, 2,862 (free ch-15), 2,941, 2,887 (free ch-100), and 2,938, 2,879 (free ch-200) in the spectra of Dox, Tam, 4-Hydroxytam, and endoxifen complexes (Fig. 4a–c). The overall spectral changes observed in this region (3,500–2,800 cm⁻¹) were attributed to the presence of both hydrophilic and hydrophobic contacts in the drug–chitosan complexes [11, 12].

172   H.A. Tajmir-Riahi et al.

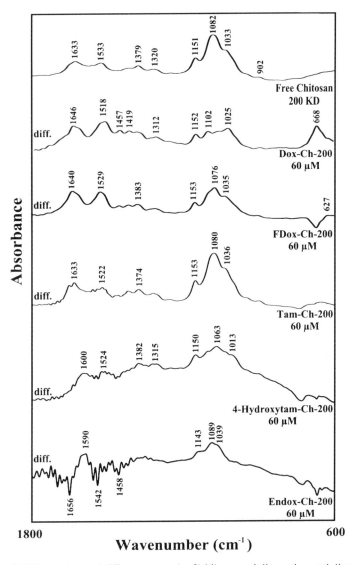

Fig. 3 FTIR spectra and difference spectra [(chitosan solution + drug solution) − (chitosan solution)] in the region of 1,800–600 cm⁻¹ for the free chitosan-200 and its drug complexes for Dox-ch-200, FDox-ch-200, tamoxifen-ch-200, 4-hydroxytamoxifen-ch-200, and endoxifen-ch-200 in aqueous solution at pH 5–6 with 60 μM drug and chitosan concentrations

2. Fluorescence quenching has been used as a convenient technique for quantifying the binding affinities of drug–polymer complexes [11, 12]. Since chitosan is a weak fluorophore, the titrations of Dox, FDox, Tam, 4-Hydroxytam, and endoxifen were done against various polymer concentrations, using drug emission bands at 350–750 nm [26, 27]. When drug interacts with chitosan, fluorescence may change depending on the impact of such interaction on the drug conformation or via direct quenching effect.

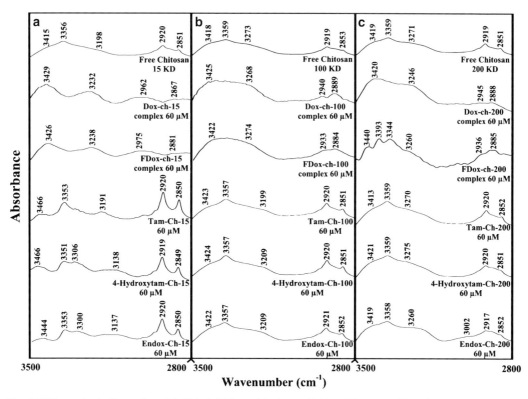

Fig. 4 FTIR spectra in the region of 3,500–2,800 cm^{-1} (polymer N–H and CH$_2$ stretching vibrations) of hydrated films (pH 5–6) for free chitosan-15 (**a**), chitosan-100 (**b**), and chitosan-200 KD (**c**) and their drug complexes obtained with 60 μM polymer and 60 μM drug concentrations

The decrease of fluorescence intensity of Dox or FDox at 592 nm and Tam, 4-Hydroxytam, and endoxifen at 365 nm was monitored for drug–chitosan systems (Figs. 5, 6, and 7). The plot of $F_0/(F_0-F)$ versus $1/$[chitosan] is shown in Figs. 5, 6, and 7 (inset). Assuming that the observed changes in fluorescence come from the interaction between the drug and the chitosan, the quenching constant can be taken as the binding constant of the complex formation. The K value given here averages four- and six-replicate run for drug–polymer systems. The overall binding constants were $K_{\text{Dox–ch-15}} = 8.4\ (\pm 0.6) \times 10^3\ \text{M}^{-1}$, $K_{\text{Dox–ch-100}} = 2.2\ (\pm 0.3) \times 10^5\ \text{M}^{-1}$, and $K_{\text{Dox–ch-200}} = 3.7\ (\pm 0.5) \times 10^4\ \text{M}^{-1}$; $K_{\text{FDox–ch-15}} = 5.5\ (\pm 0.5) \times 10^3\ \text{M}^{-1}$, $K_{\text{FDox–ch-100}} = 6.8\ (\pm 0.6) \times 10^4\ \text{M}^{-1}$, and $K_{\text{FDox–ch-200}} = 2.9\ (\pm 0.5) \times 10^4\ \text{M}^{-1}$; $K_{\text{tam–ch-15}} = 8.7\ (\pm 0.5) \times 10^3\ \text{M}^{-1}$, $K_{\text{tam–ch-100}} = 5.9\ (\pm 0.4) \times 10^5\ \text{M}^{-1}$, and $Ktam\text{–}ch\text{-}200 = 2.4\ (\pm 0.4) \times 10^5\ \text{M}^{-1}$; $K_{\text{hydroxytam–ch-15}} = 2.6\ (\pm 0.3) \times 10^4\ \text{M}^{-1}$, $K_{\text{hydroxytam–ch-100}} = 5.2\ (\pm 0.7) \times 10^6\ \text{M}^{-1}$, and $K_{\text{hydroxytam–ch-200}} = 5.1\ (\pm 0.5) \times 10^5\ \text{M}^{-1}$; and $K_{\text{endox–ch-15}} = 4.1\ (\pm 0.4) \times 10^3\ \text{M}^{-1}$, $K_{\text{endox–ch-100}} = 1.2\ (\pm 0.3) \times 10^6\ \text{M}^{-1}$, and $K_{\text{endox–ch-200}} = 4.7\ (\pm 0.5) \times 10^5\ \text{M}^{-1}$ (Figs. 5, 6, and 7 and Table 1). The order of binding constants calculated for the drug–chitosan adducts showed

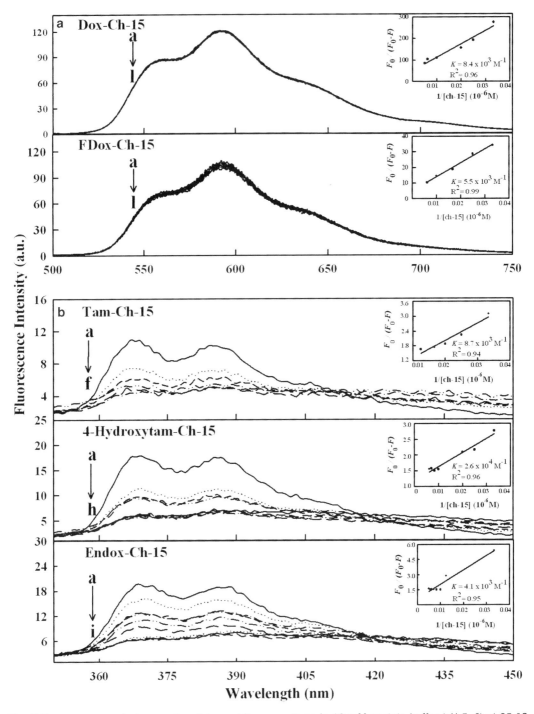

Fig. 5 Fluorescence emission spectra of drug–chitosan systems in 10 mM acetate buffer (pH 5–6) at 25 °C presented for (**a**) Dox-ch-15 and FDox and (**b**) Tam, 4-Hydroxytam, and endoxifen with free Dox (30 μM), with chitosan-15 at 10–100 μM; *inset*: K values calculated by $F_0/(F_0 - F)$ versus 1/[chitosan] for drug–chitosan-15 complexes

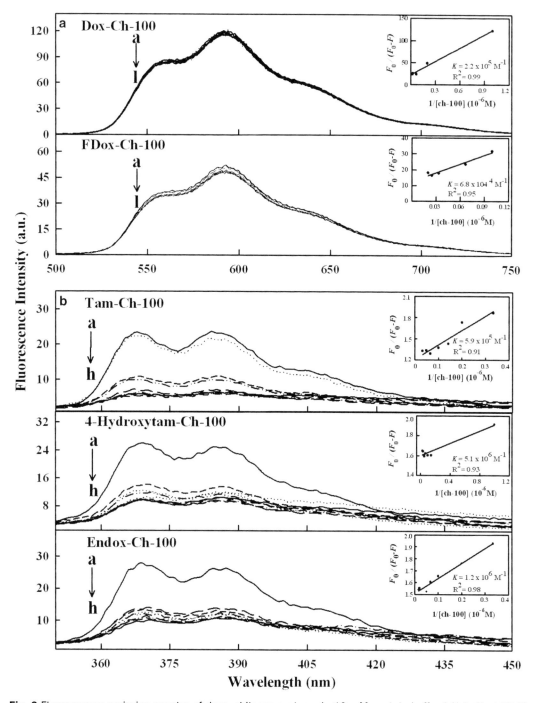

Fig. 6 Fluorescence emission spectra of drug–chitosan systems in 10 mM acetate buffer (pH 5–6) at 25 °C presented for (**a**) Dox-ch-100 and FDox and (**b**) Tam, 4-Hydroxytam, and endoxifen with free Dox (30 μM), with chitosan-100 at 10–100 μM; *inset*: K values calculated by $F_0/(F_0 - F)$ versus 1/[chitosan] for drug–chitosan-100 complexes

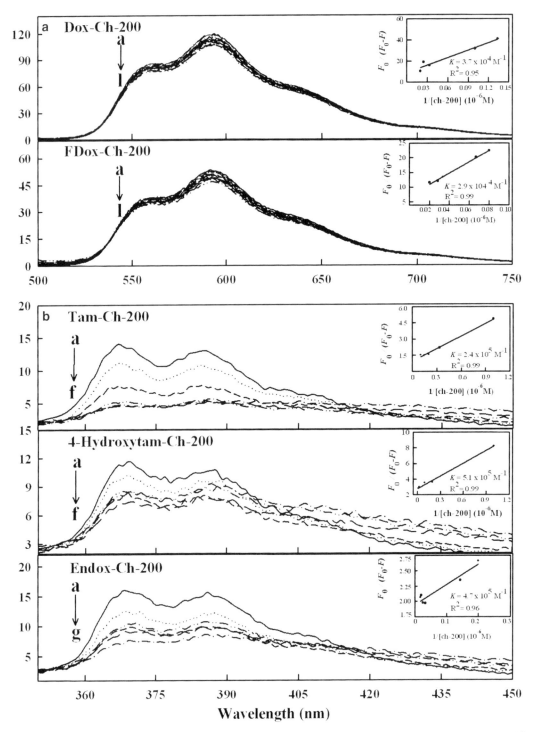

Fig. 7 Fluorescence emission spectra of drug–chitosan systems in 10 mM acetate buffer (pH 5–6) at 25 °C presented for (**a**) Dox-ch-200 and FDox and (**b**) Tam, 4-Hydroxytam, and endoxifen with free Dox (30 µM), with chitosan-200 at 10–100 µM; *inset.* K values calculated by $F_0/(F_0 - F)$ versus 1/[chitosan] for drug–chitosan-200 complexes

Table 1
Binding parameters for drug–chitosan complexes

Complex	K_{sv} (M^{-1})			K_a (M^{-1})			n			K_q (M^{-1}/s)		
	ch-15	ch-100	ch-200	ch-15	ch-100	ch-200	ch-15	ch-100	ch-200	ch-15	ch-100	ch-200
Dox	2.0×10^{10}	1.4×10^{9}	2.9×10^{8}	8.4×10^{3}	2.2×10^{5}	3.7×10^{4}	0.6	0.5	1.0	1.8×10^{19}	1.3×10^{18}	2.6×10^{17}
FDox	1.7×10^{9}	1.0×10^{9}	3.0×10^{8}	5.5×10^{3}	6.8×10^{4}	2.9×10^{4}	0.8	0.5	1.2	1.5×10^{18}	9.3×10^{17}	2.7×10^{17}
Tam	6.8×10^{7}	5.4×10^{6}	8.5×10^{7}	8.7×10^{3}	5.9×10^{5}	2.4×10^{5}	1.5	2.8	0.9	3.2×10^{16}	2.6×10^{15}	4.0×10^{16}
Hydroxytam	1.8×10^{8}	6.3×10^{7}	4.4×10^{8}	2.6×10^{4}	5.1×10^{6}	5.1×10^{5}	1.3	0.6	0.7	8.5×10^{16}	3.0×10^{16}	2.1×10^{17}
Endox	5.8×10^{7}	5.2×10^{7}	1.7×10^{8}	4.1×10^{3}	1.2×10^{6}	4.7×10^{5}	1.2	0.5	0.5	2.7×10^{16}	2.4×10^{16}	7.9×10^{16}

ch-100>200>15 KD (Table 1). It is important to note that ch-15 is smaller than ch-100 and ch-200, while drug interaction is mainly via positively charged chitosan NH_2 groups, as polymer size gets larger, more increases in overall polymer charges will result in stronger drug–polymer complexation. However, in the case of ch-200, aggregation of polymer occurs at pH near 6, which leads to lesser affinity of the aggregated ch-200 for drug interaction (self-aggregation is less observed for ch-15 and ch-100). Therefore, ch-100 forms more stable complexes than the ch-15 and ch-200 (Table 1). The results also showed stronger affinity of chitosan with tamoxifen and its metabolites than those of DOX and FDOX due to more hydrophobic character of DOX and FDOX molecules (Table 1).

The f value calculated from Eq. 7 represents the mole fraction of the accessible population of fluorophore to quencher. The f values were from 0.25 to 0.65 for these drug–chitosan complexes, indicating a large portion of fluorophore was exposed to quencher.

The number of drug binding site on polymer (n) is calculated from $\log [(F_0 - F)/F] = \log K_S + n \log [\text{chitosan}]$ for the static quenching [38–47]. The n values from the slope of the straight-line plot showed between 2.8 and 0.5 sites are occupied by drug on chitosan molecule (Fig. 8 and Table 1). The results showed some degree of cooperativity for drug–polymer interaction.

In order to verify the presence of static or dynamic quenching in drug–chitosan complexes, we have plotted F_0/F against Q to estimate the quenching constant (KQ), and the results are shown in Fig. 9. The plot of F_0/F versus Q is a straight line for drug–chitosan adducts, indicating that the quenching is mainly static in these drug–polymer complexes (Fig. 9). The quenching constant K_Q was estimated according to the Stern–Volmer equation:

$$F_0 / F = 1 + k_Q t_0 [Q] = 1 + K_{sv} [Q], \qquad (8)$$

where F_0 and F are the fluorescence intensities in the absence and presence of quencher, $[Q]$ is the quencher concentration, and K_{sv} is the Stern–Volmer quenching constant [48, 49], which can be written as $K_{sv} = k_Q t_0$, where k_Q is the bimolecular quenching rate constant and t_0 is the lifetime of the fluorophore in the absence of quencher about 1.1 ns for free Dox and FDox around neutral pH [26, 50]. The quenching constants (K_Q) are $1.8 \times 10^{19} M^{-1}/s$ for Dox-ch-15, $1.3 \times 10^{18} M^{-1}/s$ for Dox-ch-100, and $2.6 \times 10^{17} M^{-1}/s$ for Dox-ch-200; $1.5 \times 10^{18} M^{-1}/s$ for FDox-ch-15, $9.3 \times 10^{17} M^{-1}/s$ for FDox-ch-100, and $2.7 \times 10^{17} M^{-1}/s$ for FDox-ch-200; $3.2 \times 10^{16} M^{-1}/s$ for Tam-ch-15, $2.6 \times 10^{15} M^{-1}/s$ for

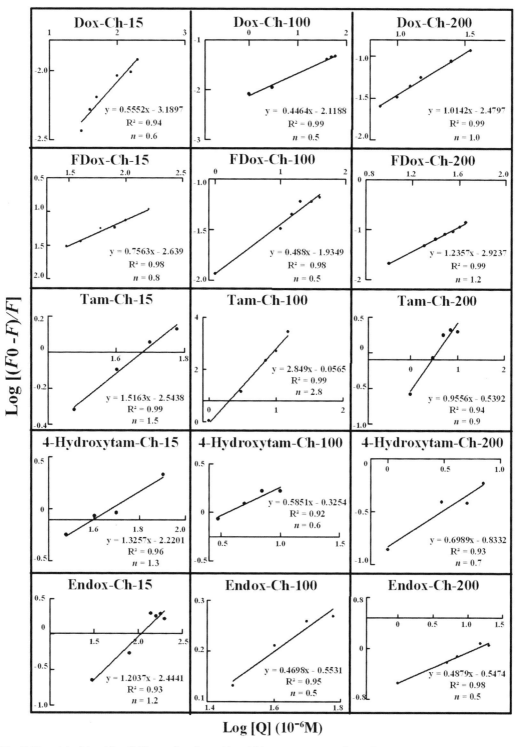

Fig. 8 The plot of log $(F_0 - F)/F$ as a function of log (chitosan concentrations) for the number of drug binding sites on chitosan (n) for drug–polymer complexes

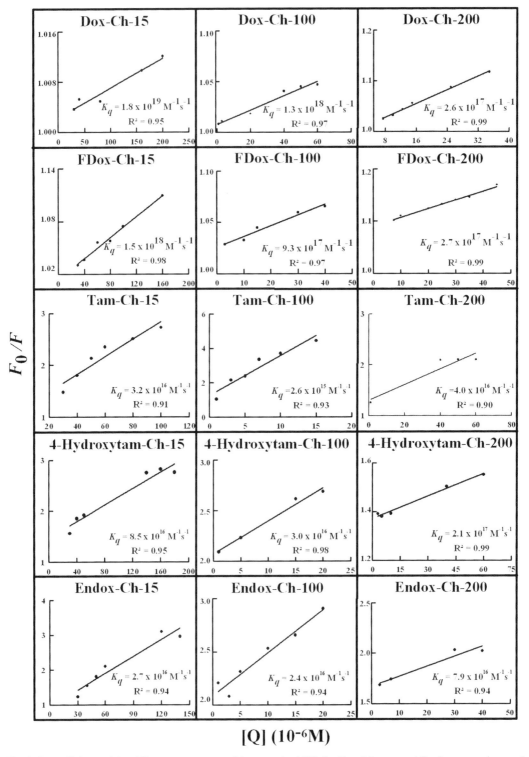

Fig. 9 Stern–Volmer plots of fluorescence quenching constant (K_q) for the chitosan and its drug complexes at different chitosan concentrations

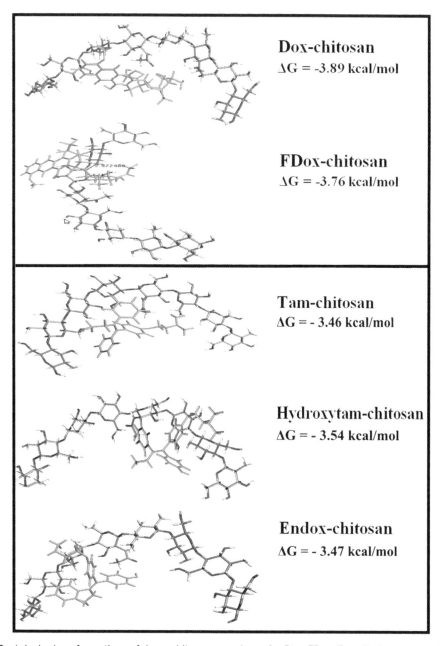

Fig. 10 Best docked conformations of drug–chitosan complexes for Dox, FDox, Tam, Hydroxytam, and endoxifen bound to chitosan with free binding energies

Tam-ch-100, and 4.0×10^{16} M^{-1}/s for Tam-ch-200; 8.5×10^{16} M^{-1}/s for 4-Hydroxytam-ch-15, 3.0×10^{16} M^{-1}/s for 4-Hydroxytam-ch-100, and 2.1×10^{17} M^{-1}/s for 4-Hydroxytam-ch-200; and 2.7×10^{16} M^{-1}/s for Endox-ch-15, 2.4×10^{16} M^{-1}/s for Endox-ch-100, and 7.9×10^{16} M^{-1}/s for Endox-ch-200 (Fig. 9 and Table 1). Since these values are much greater than the maximum collisional quenching constant $(2.0 \times 10^{10} M^{-1}/s)$, the static quenching is dominant in these drug–chitosan complexes [48].

3. Molecular modeling was often used to predict the binding sites for drug–polymer complexes. The spectroscopic results were combined with docking experiments in which Dox, FDox, Tam, 4-Hydroxytam, and endoxifen molecules were docked to chitosan to determine the preferred binding sites on the polymer. The models of the docking for drug are shown in Fig. 10. The docking results showed that drugs are surrounded by several donor atoms of chitosan C–O, N–H, and NH_2 groups on the surface with a free binding energy of –3.89 (Dox), –3.76 (FDox), –3.46 (Tam), –3.54 (4-Hydroxytam), and –3.47 kcal/mol (Fig. 10). It should be noted that FDox is located near chitosan C–O, N–H, and NH_2 groups with a hydrogen bonding system between drug O-213 and chitosan N-19 atoms $(2.922 \text{\AA})$ (Fig. 10). As one can see, drugs are not surrounded by similar donor groups showing different binding modes in these drug–chitosan complexes (Fig. 10).

5   Concluding Remarks

The spectroscopic and docking studies are strong analytical methods to determine the binding parameters and the binding sites of drug–chitosan complexes. Major hydrophilic contacts via chitosan charged NH_2 groups and hydrophobic interactions as well as H-bonding are observed in the drug–chitosan complexes. The order of drug–polymer binding is ch-100 > ch-200 > ch-15. Chitosan is stronger carrier for tamoxifen, 4-hydroxytamoxifen, and endoxifen than for doxorubicin and N-(trifluoroacetyl) doxorubicin in vitro. However, the addition of more soluble polymer such as PEG to chitosan will increase chitosan solubility and enhances drug binding affinity both in vitro and in vivo [51].

Acknowledgments

The financial support of the Natural Sciences and Engineering Research Council of Canada (NSERC) is highly appreciated.

References

1. Agudelo D, Nafisi S, Tajmir-Riahi HA (2013) Encapsulation of milk beta-lactoglobulin by chitosan nanoparticles. J Phys Chem B 117: 6403–6409

2. Amidi M, Mastrobattista E, Jiskoot W, Hennink WE (2010) Chitosan- based delivery systems for protein therapeutics and antigens. Adv Drug Deliv Rev 62:59–82

3. Gan Q, Wang T (2007) Chitosan nanoparticles as protein delivery carrier-systematic examination of fabrication conditions for efficient loading and release. Colloids Surf B: Biointerfaces 59:24–34

4. Pacheco N, Gamica-Gonzalez M, Gimeno M, Barzana E, Trombotto S, David L, Shirai K (2011) Structural characterization of chitin and chitosan obtained by biological and chemical methods. Biomacromolecules 12:3285–3290

5. Rabea EI, Badawy MET, Stevens CV, Smagghe G, Steurbaut W (2003) Chitosan as antimicrobial agent: application and mode of action. Biomacromolecules 4:1457–1465

6. Dang JM, Leong KW (2006) Natural polymers for gene delivery and tissue engineering. Adv Drug Deliv Rev 58:487–499

7. Mao S, Shuai X, Unger F, Simon M, Bi D, Kissel T (2004) The depolymerization of chitosan: effects on physicochemical and biological properties. Int J Pharm 281:45–54

8. Saranya N, Moorthi A, Saravanan S, Pandima Devi M, Selvamurugan N (2011) Chitosan and its derivatives for gene delivery. Int J Biol Macromol 49:234–238

9. Shu Z, Zhu K (2000) A novel approach to prepare tripolyphosphate:chitosan complex beads for controlled release drug delivery. Int J Pharm 201:51–58

10. Souza KS, Gonc-alves MDP, Gomez J (2011) Effect of chitosan degradation on its interaction with β-lactoglobulin. Biomacromolecules 12:1015–1029

11. Sanyakamdhorn S, Agudelo D, Tajmir-Riahi HA (2013) Encapsulation of antitumor drug doxorubicin and its analogue by chitosan nanoparticles. Biomacromolecules 14:557–563

12. Agudelo D, Sanyakamdhorn S, Nafisi S, Tajmir-Riahi HA (2013) Transporting antitumor drug tamoxifen and its metabolites, 4-hydroxytamoxifen and endoxifen by chitosan nanoparticles. PLoS ONE 8(1–11): e60250

13. Bowman K, Leong KW (2006) Chitosan nanoparticles for oral drug and gene delivery. Int J Nanomedicine 1:117–128

14. Carvalho C, Santos RX, Cardoso S, Correia S, Oliveira PJ, Santos MS, Moreira PI (2009) Doxorubicin: the good, the bad and the ugly effect. Curr Med Chem 16:3267–3285

15. Minotti G, Menna P, Salvatorelli E, Cairo G, Gianni L (2004) Anthracyclines: molecular advances and pharmacologic developments in antitumor activity and cardiotoxicity. Pharmacol Rev 56:185–229

16. Turner A, Li LC, Pilli T, Qian L, Wiley EL, Setty S, Christov K, Ganesh L, Maker AV, Li P, Kanteti P, Gupta TKD, Prabhakar BS (2013) MADD knock-down enhances doxorubicin and TRAIL induced apoptosis in breast cancer cells. PLoS ONE 8(1–8):e56817

17. Jordan VC (2006) Tamoxifen (ICI46,474) as a targeted therapy to treat and prevent breast cancer. Br J Pharmacol 147(Suppl):S269–S276

18. Spears M, Bartlett J (2009) The potential role of estrogen receptors and the SRC family as targets for the treatment of breast cancer. Expert Opin Ther Targets 13:665–674

19. Brauch J, Jordan VC (2009) Targeting of tamoxifen to enhance antitumour action for the treatment and prevention of breast cancer. Eur J Cancer 45:2274–2283

20. Jordan VC (2007) New insights into the metabolism of tamoxifen and its role in the treatment and prevention of breast cancer. Steroids 72:829–842

21. Acton EM, Tong GL (1981) Synthesis and preliminary antitumor evaluation of 5-Iminodoxorubicin. J Med Chem 24: 669–673

22. Bérubé G, Richardson VJ, Ford CHJ (1991) Synthesis of new N-(trifluoroacetyl) doxorubicin analogues. Synthetic Comm 21:931–944

23. Fauq AH, Maharvi GM, Sinha D (2010) A convenient synthesis of (Z)-4-hydroxy-N-desmethyltamoxifen (endoxifen). Bioorg Med Chem Lett 15:3036–3038

24. Brugnerotto J, Lizardi J, Goycoolea FM, Arguelles-Monal W, Desbrieres J, Rinaudo M (2001) An infrared investigation in relation with chitin and chitosan characterization. Polymer 42:3569–3580

25. Palpandi C, Shanmugam V, Shanmugam A (2009) Extraction of chitin and chitosan from shell and operculum of mangrove gastropod nerita (Dostia) crepidularia lamarcck. Intl J Med Med Sci 1:198–205

26. Beng XDB, Zhilian Y, Eccleston ME, Swartling J, Slater NKH, Kaminski CF (2008) Fluorescence intensity and lifetime imaging of

free and micellar-encapsulated doxorubicin in living cells. Nanomedicine 4:49–56

27. Engelke M, Bojarski P, Blob R, Diehl H (2001) Tamoxifen perturbs lipid bilayer order and permeability: comparison of DSC, fluorescence anisotropy, laurdan generalized polarization and carboxyfluorescein leakage studies. Biophys Chem 90:157–173

28. Dufour C, Dangles O (2005) Flavonoid-serum albumin complexation: determination of binding constants and binding sites by fluorescence spectroscopy. Biochim Biophys Acta 1721:164–173

29. Froehlich E, Jennings CJ, Sedaghat-Herati MR, Tajmir-Riahi HA (2009) Dendrimers bind human serum albumin. J Phys Chem B 113:6986–6993

30. He W, Li Y, Xue C, Hu Z, Chen X, Sheng F (2005) Effect of Chinese medicine alpinetin on the structure of human serum albumin. Bioorg Med Chem 13:1837–1845

31. Sarzehi S, Chamani J (2010) Investigation on the interaction between tamoxifen and human holo-transferrin: determination of the binding mechanism by fluorescence quenching, resonance light scattering and circular dichroism methods. Int J Biol Macromol 47:558–569

32. Bi S, Ding L, Tian Y, Song D, Zhou X, Liu X, Zhang H (2004) Investigation of the interaction between flavonoids and human serum albumin. J Mol Struct 703:37–45

33. Iranfar H, Rajabi O, Salari R, Chamani J (2012) Probing the interaction of human serum albumin with ciprofloxacin in the presence of silver nanoparticles of three sizes: multispectroscopic and ζ potential investigation. J Phys Chem B 116:1951–1964

34. Lakowicz JR (2006) In principles of fluorescence spectroscopy, 3rd edn. Springer, New York

35. Taye N, Rungassamy T, Albani JR (2009) Fluorescence spectral resolution of tryptophan residues in bovine and human serum albumins. J Pharm Biomed Anal 50:107–116

36. Skovstrip S, Hansen SG, Skrydstrup T, Schiott B (2010) Conformational flexibility of chitosan: a molecular modeling study. Biomacromolecules 11:3196–3207

37. Manikrao AM, Mahajan NS, Jawarkar DR, Khatale PN, Kedar KC, Thombare KS (2012) Docking analysis of darunnavir as HIV protease inhibitors. J Comput Meth Mol Des 2:29–43

38. Agudelo D, Beauregard M, Bérubé G, Heidar-Ali Tajmir-Riahi HA (2012) Antibiotic doxorubicin and its derivative bind milk beta-lactoglobulin. J Photochem Photobiol B 117:185–192

39. Agudelo D, Bourassa P, Bruneau J, Berubé G, Asselin E, Tajmir-Riahi HA (2012) Probing the binding sites of antibiotic drug doxorubicin and N-(trifluoroacetyl) doxorubicin with human and bovine serum albumins. PLoS ONE 7(8):1–13, e43814

40. Ahmed Belatik A, Hotchandani S, Carpentier R, Tajmir-Riahi HA (2012) Locating the binding sites of Pb(II) ion with human and bovine serum albumins. PLoS ONE 7(5):1–11, e36723

41. Mandeville JS, Tajmir- Riahi HA (2010) Complexes of dendrimers with bovine serum albumin. Biomacromolecules 11:465–472

42. Charbonneau DM, Tajmir-Riahi HA (2010) Study on the Interaction of cationic lipids with bovine serum albumin. J Phys Chem B 114:1148–1155

43. Bourassa P, Kanakis DC, Tarantilis P, Polissiou MG, Tajmir-Riahi HA (2010) Resveratrol, genistein and curcumin bind bovine serum albumin. J Phys Chem B 114:3348–3354

44. Froehlich E, Mandeville JF, Arnold D, Kreplak L, Tajmir-Riahi HA (2012) Effect of PEG and mPEG-anthracene on tRNA aggregation and article formation. Biomacromolecules 13:282–287

45. Froehlich E, Mandeville JF, Arnold D, Kreplak L, Tajmir-Riahi HA (2011) PEG and mPEG-anthracene induce DNA condensation and particle formation. J Phys Chem B 115:9873–9879

46. Mandeville JS, N'soukpoé-Kossi CN, Neault JF, Tajmir-Riahi HA (2010) Structural analysis of DNA interaction with retinol and retinoic acid. Biochem Cell Biol 88:469–477

47. Mandeville JS, Froehlich E, Tajmir-Riahi HA (2009) Study of curcumin and genistein interactions with human serum albumin. J Pharm Biomed Anal 49:468–474

48. Zhang G, Que Q, Pan J, Guo J (2008) Study of the interaction between icariin and human serum albumin by fluorescence spectroscopy. J Mol Struct 881:132–138

49. Jiang M, Xie MX, Zheng D, Liu Y, Li XY, Chen X (2004) Spectroscopic studies on the interaction of cinnamic acid and its hydroxyl derivatives with human serum albumin. J Mol Struct 692:71–80

50. Huang DC, Piché M, Ma G, Jean-Jacques M, Khayat M (2010) Early detection and treatment monitoring of human breast cancer MCF-7 using fluorescence imaging. ART Advanced Research Technologies Inc. www.art.ca/docs/publications/WMIC2010_MCF7.pdf

51. Gong C, Qi T, Wei X et al (2013) Thermosensitive polymeric hydrogels as drug delivery systems. Curr Med Chem 20:79–94

Chapter 12

A Method for Evaluating Nanoparticle Transport Through the Blood–Brain Barrier In Vitro

Daniela Guarnieri, Ornella Muscetti, and Paolo A. Netti

Abstract

Blood–brain barrier (BBB) represents a formidable barrier for many therapeutic drugs to enter the brain tissue. The development of new strategies for enhancing drug delivery to the brain is of great importance in diagnostics and therapeutics of central nervous system (CNS) diseases. In this context, nanoparticles are an emerging class of drug delivery systems that can be easily tailored to deliver drugs to various compartments of the body, including the brain. To identify, characterize, and validate novel nanoparticles applicable to brain delivery, in vitro BBB model systems have been developed. In this work, we describe a method to screen nanoparticles with variable size and surface functionalization in order to define the physicochemical characteristics underlying the design of nanoparticles that are able to efficiently cross the BBB.

Key words Blood–brain barrier, Brain capillary endothelial cells, Nanoparticles, Transcytosis, Nanoparticle surface functionalization, Nanoparticle size

1 Introduction

The prevalence of central nervous system (CNS) diseases is likely to increase in the future due to an aging population requiring the development of improved therapies. However, despite significant efforts, treatments remain limited due to the inability of therapeutic agents to effectively cross the blood–brain barrier (BBB), which is a dynamic barrier protecting the brain against invading organisms and unwanted substances [1]. The formation of tight junctions between the endothelial cells and the presence of supporting cells (astrocytes, pericytes, and microglia), enzymes, receptors, transporters, and efflux pumps control and limit the access of molecules to the brain, making the BBB a formidable obstacle to the effective delivery of drugs to the CNS [2].

Therefore, delivering therapeutic drugs to the CNS effectively, safely, and conveniently is becoming more important than ever. Most strategies to transport pharmaceutical compounds into the

Kewal K. Jain (ed.), *Drug Delivery System*, Methods in Molecular Biology, vol. 1141,
DOI 10.1007/978-1-4939-0363-4_12, © Springer Science+Business Media New York 2014

CNS cause disruption in the anatomical texture of the BBB, therefore impairing its natural function [2, 3]. On the other hand, the biomedical and pharmaceutical applications of nanotechnology have greatly facilitated the diagnosis and treatment of CNS diseases. A number of nanoparticulate delivery systems with promising properties have been developed [4]. In particular, targeted delivery of a therapeutic cargo to the intended site of action in the brain appears to be one of the most promising noninvasive approaches to overcome the BBB, combining the advantages of brain targeting, high incorporation capacity, reduction of side effects, and circumvention of the multidrug efflux system [5–8].

Several transport mechanisms across the BBB have been identified, including paracellular or transcellular pathways, transport proteins, receptor-mediated transcytosis, and adsorptive transcytosis [2]. Many studies suggest nanoparticles bind to functional molecules, like apolipoprotein [9, 10] or transferrin [11], enabling strategies to utilize existing pathways for accessing the brain. However, the interactions of nanoparticles at the BBB have not yet been thoroughly described, although interest in this arena is rapidly increasing [12–14].

Several in vitro studies have been developed in order to address some of the fundamental mechanisms involved in crossing of BBB by solutes, drugs, and nanoparticles [15, 16]. Except their comparatively high permeability and the loss of expression of some BBB efflux protein systems [17], in vitro models have several advantages over in vivo models, including low-cost, high-throughput screening and easiness to assess compounds and to investigate the transport mechanism at the molecular levels.

In this work, we describe in details the methods used to investigate the effect of nanoparticle characteristics on the interaction with the BBB to rank them in order of efficiency. In particular, polystyrene nanoparticles with variable size and surface charge have been tested with bEnd.3 cell monolayer, grown on a Transwell system, in order to identify the parameters regulating the nanoparticle/BBB interaction and, thus, the nanoparticle's capability to pass through the endothelial barrier. This study paves the way to elucidate the mechanisms utilized by nanoparticles to cross the BBB in order to design safe and effective nanoparticles for novel diagnostic and therapeutic applications for CNS disorders.

2   Materials

2.1   Polystyrene Nanoparticles

1. Nanoparticles used in this work are the following:

   (a) Green and red fluorescent polystyrene nanoparticles with diameters, respectively, of 44 (NP44) and 100 (NP100) nm (Duke Scientific Corporation).

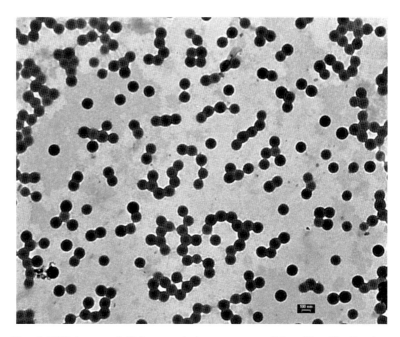

Fig. 1 TEM image of 100 nm polystyrene nanoparticles. Magnification bar: 100 nm

    (b) Yellow–green fluorescent carboxylate-modified polystyrene nanoparticles, 100 nm (NP-COOH) (Life Technologies, USA).

    (c) Orange fluorescent amine-modified polystyrene nanoparticles, 100 nm (NP-NH$_2$) (Sigma-Aldrich).

2. Morphology of polystyrene nanoparticles is verified by transmission electron microscopy (TEM) (Fig. 1).

3. To elucidate the colloidal stability of nanoparticles, measurements of the zeta-potential and of the hydrodynamic diameter (D_H) are carried out with a Zetasizer Nano ZS (Malvern Instruments, Worcestershire, UK) at 25 °C by using a nanoparticle concentration, corresponding to 0.9×10^{10} NP/ml. At least 3–5 measurements of all samples should be performed to have a statistical significance (*see* **Note 1**). The size and zeta-potential of nanoparticles used in this work are reported in Table 1.

2.2 Cell Culture

1. Immortalized mouse cerebral endothelial cells, bEnd.3 cells (American Type Culture Collection, Manassas, VA), are grown in Dulbecco's Modified Eagle's Medium (DMEM) with 4.5 g/l glucose, 10 % fetal bovine serum (FBS) (Gibco), 4 mM glutamine (Gibco), 100 U/ml penicillin, and 0.1 mg/ml streptomycin (Gibco) in a 100-mm-diameter cell culture dish (Corning Incorporated, Corning, NY) in a humidified atmosphere at 37 °C and 5 % CO$_2$.

Table 1
Size and ζ-potential measurements of nanoparticles used in this work

	NP44	NP100	NP-COOH	NP-NH2
Size [nm][a]	43.67	106.70	110.32	102.46
SD [±]	1.08	5.35	0.12	0.13
ζ-potential [mV]	−25.25	−21.97	−40.42	+41.72
SD [±]	5.26	2.11	1.39	2.18

[a]PDI < 0.1

Table 2
Transendothelial electrical resistance of bEnd.3 cells cultured on Transwell inserts up to 12 days of culture

Days of culture	3	4	7	10	12
TEER [Ohm/cm²]	122.5	338	379.33	140.5	120.33
SD [±]		3.54	17.98	19.60	6.36	8.08

2. Cell culture medium is changed every 3–4 days, and cells are split after reaching confluency.

3. Phosphate-buffered saline (PBS) solution and trypsin/EDTA solution are from Gibco.

4. Cells used in all experiments are at passages 28–35.

5. To evaluate nanoparticle cellular uptake, 5×10^4 cells are seeded in a 96-well plate.

6. For permeability experiments, cells were seeded on Transwell permeable inserts as described below.

2.3 TEER Measurements

Tight junction formation and BBB functionality are assessed by transendothelial electrical resistance (TEER) across the filters using an electrical resistance system (ERS) with a current-passing and voltage-measuring electrode (Millicell-ERS, Millipore Corporation, Bedford, MA). TEER ($\Omega \cdot cm^2$) is calculated from the displayed electrical resistance on the readout screen by subtraction of the electrical resistance of a filter without cells and a correction for filter surface area (Table 2).

2.4 Permeability Experiments

1. Seed 3×10^4 cells in 150 μl of complete medium on Transwell permeable inserts (6.5 mm in diameter, 3 μm pore size) (Corning Incorporated, Corning, NY), add 400 μl of complete medium in the basal compartment of the Transwell system, and allow them to grow up to 7 days (see **Note 2**). Change cell culture medium every 3–4 days.

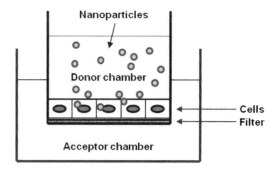

Fig. 2 Schematic representation of the Transwell system. The apical compartment (*donor chamber*) represents the blood side where nanoparticles are dispersed. The basolateral compartment (*acceptor chamber*) represents the brain side

2. Dilute nanoparticles in complete cell culture medium w/o phenol red at the final concentration of 0.002 % solids, corresponding to 4.2×10^{11} for 44 nm NPs and 3.6×10^{10} NP/ml for NP100, NP-COOH, and NP-NH$_2$, and sonicate NP suspension for 3 min prior to use.

3. A schematic representation of the system used for permeability experiments is reported in Fig. 2.

2.4.1 Fluorescent Probes

1. Albumin from bovine serum (BSA), Alexa Fluor® 488 conjugate (Life Technologies, USA), is used as a standard fluorescent probe for permeability measurements.

2. Working solutions of fluorescent probes are prepared by dissolving the BSA in complete medium w/o phenol red at the final concentration of 0.5 mg/ml (*see* **Note 3**).

3. Solution is sonicated for 2–3 min to promote BSA dissolution.

2.5 Indirect Immunofluorescence

1. 4 % paraformaldehyde solution is prepared by dissolving paraformaldehyde powder in PBS at about 100 °C under stirring. When the solution becomes transparent, make aliquots of about 4–5 ml in 15 ml conic tubes and store them at −20 °C (*see* **Note 4**).

2. 0.1 % Triton X-100 solution is prepared by diluting Triton X-100 (Sigma) in PBS under stirring. Store the solution at room temperature.

3. Prepare PBS–BSA solution by dissolving 0.5 % (w/v) bovine serum albumin (BSA) (Sigma) in PBS at room temperature (*see* **Note 5**).

4. Tight junctions are localized by incubating samples first with mouse anti-claudin-5 primary antibodies (Invitrogen, Life Technologies) diluted 1:100 in PBS–BSA 0.5 % and then with Alexa Fluor 488 anti-mouse secondary antibodies diluted 1:500 in PBS–BSA 0.5 % (Invitrogen, Life Technologies).

5. Rabbit anti-EEA1 primary antibodies (ABR) diluted 1:200 in PBS–BSA 0.5 % are used with Alexa Fluor goat anti-rabbit secondary antibodies diluted 1:500 in PBS–BSA 0.5 % (Invitrogen, Life Technologies) to label early endosomes.

6. Caveolae are localized by incubating samples first with rabbit anti-caveolin 1 (Abcam) primary antibodies diluted 1:200 in PBS–BSA 0.5 % and then with Alexa Fluor goat anti-rabbit secondary antibodies diluted 1:500 in PBS–BSA 0.5 % (Invitrogen, Life Technologies).

7. The excitation wavelengths of secondary antibodies are 488 and 568 nm for samples treated with red and green fluorescent nanoparticles, respectively.

8. Lysosomes are localized by using LysoTracker (Invitrogen, Life Technologies) diluted 1:13,000 in cell culture medium and incubating cells 30 min at 37 °C with LysoTracker solution.

9. Cell membranes are localized by using wheat germ agglutinin (WGA), Alexa Fluor 488 conjugate (Invitrogen, Life Technologies). Dilute 1 mg/ml WGA stock solution 1:200 into PBS.

2.6 Transmission Electron Microscopy (TEM)

1. Dilute glutaraldehyde (Sigma, Germany) in sodium cacodylate buffer 0.1 M (pH 7.2) (Electron Microscopy Sciences, USA) containing 1 % saccharose (Electron Microscopy Sciences, USA) at the final concentration of 2.5 % v/v.

2. Dilute aqueous osmium tetroxide (Electron Microscopy Sciences, USA) in sodium cacodylate buffer 0.1 M (pH 7.2) at the final concentration of 1 % v/v.

3. Prepare a graded series of ethanol from 30 to 100 % by diluting absolute ethanol (Sigma, Germany) in Milli-Q water.

4. EPON EMbed 812 resin is purchased by Electron Microscopy Sciences, USA.

5. Dissolve uranyl acetate (Merck, Germany) in methanol/Milli-Q water (1:1) at the final concentration of 1 % w/v.

6. Reynolds lead citrate (Merck, Germany).

3 Methods

3.1 Nanoparticle Uptake Mechanisms

Indirect immunofluorescence against endocytic markers may give information about the endocytic mechanisms underlying nanoparticle cellular uptake (Fig. 3). The co-localization experiments are carried out on 70–80 % confluent cells seeded on 12-mm-diameter glass coverslips. All the reagents and solutions are prepared as described above:

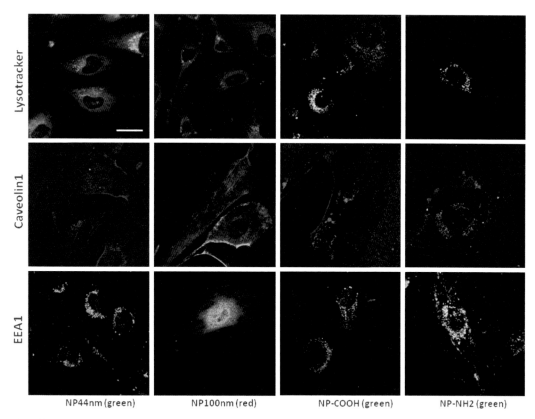

Fig. 3 Co-localization of nanoparticles with endocytic markers. Confocal images indicate a partial co-localization of polystyrene nanoparticles with early endosomes, lysosomes, and caveolae after 24 h incubation with bEnd.3 cells. Magnification bar: 20 µm

1. After nanoparticle incubation, first rinse cells twice with 1 ml PBS to remove non-internalized NPs and fix them with 400 µl of 4 % paraformaldehyde for 20 min at room temperature (*see* **Note 6**).

2. Permeabilize the cells with 1 ml 0.1 % Triton X-100 in PBS for 10 min at room temperature.

3. Wash twice the samples with 1 ml PBS.

4. Incubate the cells with 1 ml 0.5 % BSA in PBS for 15 min at room temperature to block unspecific sites.

5. Incubate the cells with 50 µl primary antibodies, diluted in PBS–BSA 0.5 % as described above, for 1 h at room temperature in a humidified chamber (*see* **Note 7**).

6. Afterward, rinse the sample three times with 1 ml PBS–BSA 0.5 % for 10 min at room temperature to remove excess of antibodies.

7. Incubate the cells with 50 µl secondary antibodies, diluted in PBS–BSA 0.5 % as described above, for 1 h at room temperature in a humidified chamber.

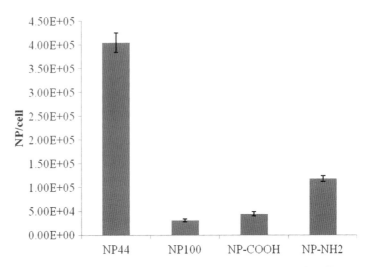

Fig. 4 Quantification of nanoparticle uptake after 24 h incubation with bEnd.3 cells

8. Afterward, rinse the sample three times with 1 ml PBS–BSA 0.5 % for 10 min at room temperature to remove excess of antibodies.

9. Mount the coverslips on glass slides with PBS/glycerol (1:1) solution.

10. Samples are observed by a confocal laser scanning microscope (CLSM) (Zeiss, Germany), equipped with an argon laser, at a wavelength of 488 nm and a He–Ne laser at a wavelength of 543 nm and 63× objectives. Images are acquired with a resolution of 1024 × 1024 pixels. The emitted fluorescence is detected using filters LP 505, BP 560–600 for red fluorescence and HFT 488/543 for FITC.

3.2 Quantification of Nanoparticle Uptake

1. To evaluate cellular uptake of nanoparticles, as a function of nanoparticle size and surface functionalization, cells are incubated at 37 °C with nanoparticle suspensions at the final concentrations described above.

2. After 24 h incubation, cells are rinsed with PBS and lysed with 100 μl of 1 % Triton X-100 in PBS.

3. Cell lysates are analyzed by a spectrofluorometer (Wallac 1420 VICTOR2, PerkinElmer, USA) to measure the fluorescence intensity of internalized nanoparticles.

4. Nanoparticle concentration in the samples is determined by interpolating the fluorescence intensity data of the samples with the calibration curve (*see* **Note 8**) and normalized to cell number (Fig. 4).

3.3 Nanoparticle Transcytosis Experiments

The permeability experiments are performed on the monolayer 7 days after cell seeding, allowing sufficient time for the cells to develop the tight junctions (Fig. 5 and Table 2) (*see* **Note 9**):

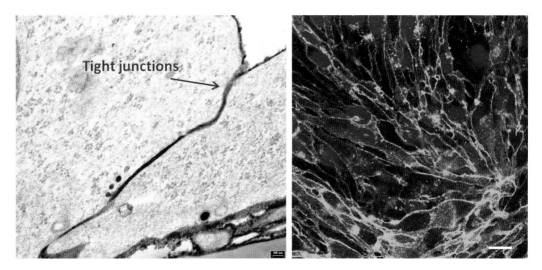

Fig. 5 Tight junction (TJ) formation in bEnd.3 monolayer after 7 days of culture on Transwell filter by TEM analysis (*left panel*) (magnification bar: 100 nm). Indirect immunofluorescence against claudin-5 protein (*green*) indicating the organization of TJs and DAPI staining showing cell nuclei (*blue*) (*right panel*) (magnification bar: 20 μm)

1. On the day of experiment, prepare the nanoparticle suspension by diluting nanoparticle stock in complete media w/o phenol red as described above.

2. Remove the media and wash the Transwell insert filter where cells are seeded with PBS with Ca^{2+} and Mg^{2+} (Gibco) to eliminate traces of phenol red that could affect fluorescence measurements.

3. Then, fill first the donor chamber with 0.15 ml complete medium w/o phenol red containing nanoparticles and second the acceptor chamber with 0.4 ml complete medium w/o phenol red (Fig. 2) (*see* **Note 10**).

4. Collect the samples of 0.4 ml every 10 min for 90 min from the acceptor chamber and then replace them with the same amount of complete medium w/o phenol red.

5. Measure the fluorescence intensity of the samples by spectrophotometer. The excitation and emission wavelengths are set to 485 and 535 nm, respectively, for all fluorescence tracers in the present study (*see* **Note 11**).

6. At the end of 90 min, wash the filter and fill first the donor chamber with 0.15 ml complete medium without phenol red containing 0.5 mg/ml BSA and second the acceptor chamber with 0.4 ml complete medium w/o PR.

7. Repeat **steps 4** and **5** (*see* **Note 12**).

8. To have a statistical significance, the experiments should be performed in triplicate at least.

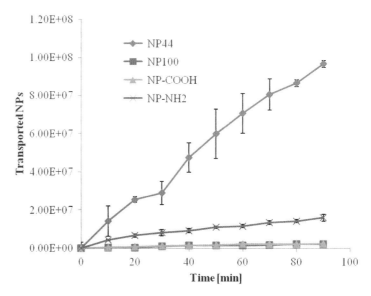

Fig. 6 Transport of NP44, NP100, NP-NH$_2$, and NP-COOH across the BBB layer. Data indicate a higher transport of smaller (NP44) and positively charged (NP-NH2) nanoparticles compared to NP100 and NP-COOH

3.3.1 Calculation of Nanoparticle Concentration in the Acceptor Chamber

Nanoparticle concentration in the samples is determined by interpolating the fluorescence intensity data of the samples with the calibration curve:

1. For the calibration curve, prepare a 1:2 dilution series of nanoparticle standards in the concentration range of $0–3.6 \times 10^{10}$ NP/ml by diluting the nanoparticle stocks in cell culture media w/o phenol red.

2. Sonicate nanoparticle suspensions for 5 min at room temperature in ultrasonic bath.

3. Transfer 100 μl of each nanoparticle standard to a 96-well plate suitable for fluorescence measurement.

4. Read the fluorescence with a fluorescence plate reader at 488 nm excitation/530 nm emission for green and orange fluorescent nanoparticles and at 543 nm excitation/560 nm emission for red fluorescent nanoparticles.

5. Transport of nanoparticles across the BBB depends on nanoparticle size and surface charge. In particular, smaller (NP44) and positively charged (NP-NH$_2$) nanoparticles are transported more efficiently through the endothelial layer than NP100 and NP-COOH (Fig. 6) (*see* **Note 13**).

3.3.2 Calculation of Permeability

The solute permeability (*P*) of the monolayer is calculated from the relationship:

$$P = \frac{\dfrac{\Delta C_{\mathrm{A}}}{\Delta t} \times V_{\mathrm{A}}}{C_{\mathrm{D}} \times S},$$

Table 3
Permeability values of fluorescent BSA after NP exposure compared to control cells

	BSA (control)	BSA (NP44 nm)	BSA (NP100 nm)	BSA (NP-COOH)	BSA (NP-NH2)
Permeability [$\times 10^{-6}$ cm/s]	3.44	3.00	2.94	2.50	3.62
SD [$\pm$]	0.30	0.16	0.42	0.43	0.02

where $\Delta C_A / \Delta t$ is the increase in fluorescence concentration in the acceptor chamber during the time interval Δt, C_D is the fluorescence concentration in the donor chamber (assumed to be constant during the experiment), V_A is the volume of the acceptor chamber, and S is the surface area of the filter (*see* **Note 14**).

P values of BSA in control cells and in cells after nanoparticle exposure are reported in Table 3.

3.4 Confocal Microscopy Analysis of Transwell Filters After Permeability Experiments

Confocal microscopy observations are carried out to verify the integrity of the endothelial layer and the presence of nanoparticles within the cells (Fig. 7, left panel):

1. After 90 min incubation with nanoparticles, wash twice the filter with PBS to remove excess of nanoparticles.

2. Fix the cells with 4 % paraformaldehyde in PBS for 20 min at room temperature.

3. After fixation, wash the samples with PBS.

4. Incubate the cells with fluorescent WGA solution in PBS for 10 min at room temperature to stain cell membranes.

5. After incubation, wash twice the samples with PBS.

6. Cut the filter from the Transwell support with a scalpel and wash it with Milli-Q water.

7. Remove excess of water by gently drying the filter with paper.

8. Mount the filter on a glass slide with PBS/glycerol (1:1) solution and cover it with a glass coverslip (*see* **Note 15**).

9. Observe the samples by the confocal microscope.

3.5 Transmission Electron Microscopy Analysis of Transwell Filters After Permeability Experiments

TEM is performed to precisely localize the intracellular particles and to confirm the nanoparticle transport via a transcellular process (Fig. 7, right panel):

1. After 90 min (see above) incubation with nanoparticles, wash twice the filter with PBS to remove excess of nanoparticles.

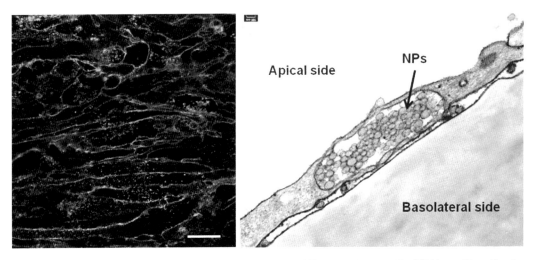

Fig. 7 CLSM (*left panel*) and TEM (*right panel*) images of 100 nm NP transport across the BBB layer. Magnification bar: 20 μm (*left panel*) and 100 nm (*right panel*)

2. Fix first the cells with 2.5 % glutaraldehyde (Sigma, Germany) in sodium cacodylate buffer 0.1 M (pH 7.2) containing 1 % saccharose (Sigma, Germany) for 2 h at room temperature.

3. Wash the samples three times for 10 min with sodium cacodylate buffer and then fix with 1 % aqueous osmium tetroxide (Electron Microscopy Sciences, USA) for 1 h at 4 °C.

4. Wash the samples three times for 10 min with sodium cacodylate buffer.

5. Afterward, dehydrate the samples in a graded series of ethanol (Sigma, Germany) for 15 min at 4 °C from 30 to 90 % ethanol.

6. Dehydrate the samples for 15 min in 100 % ethanol at room temperature.

7. Repeat the dehydration in 100 % ethanol three times.

8. Embed the samples in EPON and polymerize at 60 °C.

9. Prepare ultrathin sections (70 nm thickness) by ultramicrotome (Reichert-Jung Ultracut E) and collect them on copper grids (200 mesh).

10. The prepared ultrathin sections are contrasted with Reynolds lead citrate and with 1 % uranyl acetate.

11. Observe the samples with a transmission electron microscope (Philips EM 208 S) at an accelerating voltage of 80 kV.

4   Notes

1. Repeat size measurements before and after 10 min sonication in ultrasound bath in order to verify any influence of sonication on nanoparticle size.

2. Filter pore size must be bigger than nanoparticle size in order to avoid filter hindrance to nanoparticle passage. For nanoparticles with a diameter ≥100 nm, a 3.0-µm-pore-size filter is recommended.

3. Prepare aliquots of fluorescent BSA solution and store them at −20 °C. Avoid freezing and thawing the aliquots repeatedly.

4. Thaw 4 % paraformaldehyde in a thermostatic bath at 37 °C prior to use and maintain it at 4 °C. Do not refreeze.

5. It is recommended to prepare the PBS–BSA solution at the time of use. To store it for longer time, add NaN_3 at the final concentration of 0.02 %.

6. Take care to completely cover the cells with the paraformaldehyde solution.

7. Prepare a humidified chamber by putting wet paper in a plastic box with an airtight lid in order to avoid evaporation of the antibody solution.

8. For the calibration curve, prepare a 1:2 dilution series of nanoparticle standards in the concentration range of $0–3.6 \times 10^{10}$ NP/ml by diluting the nanoparticle stocks in cell lysis buffer. Transfer 100 µl of each nanoparticle standard to a 96-well plate suitable for fluorescence measurement. Read the fluorescence with a fluorescence plate reader at 488 nm excitation/530 nm emission for green and orange fluorescent nanoparticles and at 543 nm excitation/560 nm emission for red fluorescent nanoparticles.

9. Verify possible interactions between the filter and the nanoparticles by carrying out the permeability experiments using the Transwell system without cells as control. Afterward, cut the filter from the support and observe it at the fluorescence microscope to verify the presence of nanoparticles attached to it.

10. This is a critical step of the procedure. In this phase, please take care not to drop the media of the donor chamber into the acceptor chamber. Moreover, in order to avoid the floating of the filter that can perturb the endothelial layer integrity, fill first the donor chamber and then the acceptor one.

11. The presence of nanoparticle aggregates could affect the fluorescence measurements. Therefore, it is recommended to sonicate the samples before fluorescence measurements in order to avoid inaccuracy of the measurement due to the presence of NP aggregates in the media.

12. To verify the integrity of the endothelial barrier after nanoparticle exposure, it is important to test the permeability of the cell layer to a fluorescently labeled standard molecule, such as BSA or 70 kDa dextran. These steps allow the evaluation of

the functionality of the BBB layer after nanoparticle exposure. It is recommended to use BSA labeled with a fluorophore having a different excitation wavelength from nanoparticles. Compare the permeability data of the samples with control cells (not treated with nanoparticles). TEER measurements should be avoided because, after NP exposure, a general increment in *trans*endothelial resistance values occurs due to the cellular internalization of NPs.

13. Although the BBB in vitro systems do not represent exactly what happens into the brain, we feel that they are able to rank in order of efficiency in terms of nanoparticle characteristics (e.g., size, surface charge, shape, etc.) and to investigate the mechanisms utilized by nanoparticles to cross the BBB. In this respect, it is of relevance that the BBB in vitro system has been used also for screening nanoparticles functionalized with a cell-penetrating peptide (gH625) deriving from herpes simplex virus type 1 and showing a particular tropism for cell membrane lipids. Results indicate an enhancement of BBB crossing due to nanoparticle surface functionalization [18].

14. Transwell insert itself could also provide resistance on transport of solutes. Therefore, the total resistance of this in vitro model system comprises of two parts: the resistance from the monolayer and the resistance from the empty insert. The permeability of the cell monolayer (P_c) is calculated by the following equation:

$$P_c = \frac{P_t}{1 - \dfrac{P_t}{P_i}}.$$

Here P_t is the measured permeability of the total system and P_i is the permeability of the insert.

15. Take care to put the upper/cellular side of the filter in contact with the coverslip and the bottom side with the glass slide, in order to improve the visualization of the cells by confocal microscope. Moreover, to avoid wrinkles, cut smaller pieces of the filter with the scalpel.

Acknowledgements

The authors would like to acknowledge Dr. Valentina Mollo for her excellent experimental contribution to the data presented in this paper.

References

1. Neuwelt E, Abbott N, Abrey L et al (2008) Strategies to advance translational research into brain barriers. Lancet Neurol 7:84–96
2. Abbott NJ, Ronnback L, Hansson E (2006) Astrocyte-endothelial interactions at the blood–brain barrier. Nat Rev Neurosci 7:41–53
3. Tosi G, Costantino L, Rivasi F et al (2007) Targeting the central nervous system: in vivo experiments with peptide-derivatized nanoparticles loaded with loperamide and rhodamine-123. J Control Release 122:1–9
4. Yang H (2010) Nanoparticle-mediated brain-specific drug delivery, imaging, and diagnosis. Pharm Res 27:1759–1771
5. Mahajan SD, Law W-C, Aalinkeel R, Reynolds J, Nair BB, Yong K-T, Roy I, Prasad PN, Schwartz SA (2012) Nanoparticle-mediated targeted delivery of antiretrovirals to the brain, In Nanomedicine: Infectious Diseases, Immunotherapy, Diagnostics, Antifibrotics, Toxicology and Gene Medicine (Duzgunes, N., Ed.), pp 41–60
6. Jain KK (2012) Nanobiotechnology-based strategies for crossing the blood–brain barrier. Nanomedicine 7:1225–1233
7. Orive G, Ali OA, Anitua E, Pedraz JL, Emerich DF (2010) Biomaterial-based technologies for brain anti-cancer therapeutics and imaging. Biochim Biophys Acta 1806:96–107
8. Qiao R, Jia Q, Huewel S et al (2012) Receptor-mediated delivery of magnetic nanoparticles across the blood–brain barrier. ACS Nano 6:3304–3310
9. Cedervall T, Lynch I, Foy M et al (2007) Detailed identification of plasma proteins adsorbed on copolymer nanoparticles. Angew Chem Int Ed Engl 46:5754–5756
10. Hellstrand E, Lynch I, Andersson A et al (2009) Complete high-density lipoproteins in nanoparticle corona. FEBS J 276: 3372–3381
11. van Rooy I, Mastrobattista E, Storm G, Hennink WE, Schiffelers RM (2011) Comparison of five different targeting ligands to enhance accumulation of liposomes into the brain. J Control Release 150:30–36
12. Oberdoerster G, Elder A, Rinderknecht A (2009) Nanoparticles and the brain: cause for concern? J Nanosci Nanotechnol 9: 4996–5007
13. Bhaskar S, Tian F, Stoeger T et al (2010) Multifunctional nanocarriers for diagnostics, drug delivery and targeted treatment across blood-brain barrier: perspectives on tracking and neuroimaging. Part Fibre Toxicol 7
14. Zensi A, Begley D, Pontikis C et al (2010) Human serum albumin nanoparticles modified with apolipoprotein A-I cross the blood-brain barrier and enter the rodent brain. J Drug Target 18:842–848
15. Ragnaill MN, Brown M, Ye D et al (2011) Internal benchmarking of a human blood-brain barrier cell model for screening of nanoparticle uptake and transcytosis. Eur J Pharm Biopharm 77:360–367
16. Yuan W, Li G, Fu BM (2010) Effect of surface charge of immortalized mouse cerebral endothelial cell monolayer on transport of charged solutes. Ann Biomed Eng 38: 1463–1472
17. Nicolazzo JA, Charman SA, Charman WN (2006) Methods to assess drug permeability across the blood–brain barrier. J Pharm Pharmacol 58:281–293
18. Guarnieri D, Falanga A, Muscetti O et al (2013) Shuttle-mediated nanoparticle delivery to the blood–brain barrier. Small 9: 853–862

Chapter 13

Bacterial Systems for Gene Delivery to Systemic Tumors

Joanne Cummins, Michelle Cronin, Jan Peter van Pijkeren, Cormac G.M. Gahan, and Mark Tangney

Abstract

Certain bacteria have emerged as biological gene vectors with natural tumor specificity, capable of specifically delivering genes or gene products to the tumor environment when intravenously (i.v.) administered to rodent models. Here, we describe procedures for studying this phenomenon in vitro and in vivo for both invasive and noninvasive bacteria suitable for exploitation as tumor-specific therapeutic delivery vehicles, due to their ability to replicate specifically within tumors and/or mediate bacterial-mediated transfer of plasmid DNA to mammalian cells (bactofection).

Key words Mouse models, In vitro gene delivery models, In vivo gene delivery models

1 Introduction

Since the late 1800s, there have been numerous cases of bacteria detected within various tumors [1–3]. The mechanism of how bacteria replicate and survive within tumors includes several different elements. The chief means by which this occurs are believed to be due to the hypoxic nature of many solid tumors; this results in low oxygen levels compared to normal tissues, providing distinctive growth conditions for both anaerobic and facultative anaerobic bacteria [4]. Additional aspects influencing bacterial replication within the tumor include the presence of bacterial nutrients within the necrotic zone such as purines [5, 6]. Also, the participation of bacterial chemotaxis towards chemoattractant compounds located in necrotic areas (e.g., aspartate, serine, citrate, ribose, or galactose) produced by dormant cancer cells has also been proposed as a causative component [5, 6]. Recent findings in this area have stated that other key aspects for tumor-specific bacterial replication include atypical neovasculature and local immune suppression [5, 6]. As tumors mature, they stimulate the formation of new blood vessels (neo-angiogenesis). Yet these newly fashioned vessels are highly disorganized with unfinished endothelial linings and blind ends, subsequently resulting in "leaky" blood vessels and slow blood flow. This permeable tumor vasculature may allow circulating bacteria to enter tumor tissue and entrench locally [5, 6]. In addition, an assortment of devices are utilized by cancerous cells to evade detection by the immune system ensuing inadequate immune response within tumors, possibly allowing a sanctuary for bacteria to elude immune clearance, which is not found elsewhere in the body [7, 8].

The ideal anticancer treatment should distinguish between tumor tissue and normal tissue. The use of bacteria as therapeutic agents to target tumors presents an appealing tactic, since we and others have shown in tumor mouse models that bacteria naturally replicate within tumors when systemically administered [9, 10]. In the case of noninvasive species, strains can be engineered to secrete therapeutic proteins restricted to the tumor environment, peripheral to tumor cells. This cell therapy approach is especially suitable for indirectly acting therapeutic approaches, such as antiangiogenesis and immune therapy [5]. Gene-based anti-angiogenic therapy has been used in combination with other applications to decrease angiogenesis. An example of such therapies includes bifidobacterial expression of endostatin genes, an endogenous inhibitor of angiogenesis, which have shown potential in preclinical trials [5].

Bacterial studies have principally concentrated on the delivery of genes for ensuing tumor cell expression of anticancer agents exploiting pathogenic invasive bacterial species. Bactofection is a technique using bacteria for direct gene transfer into the target organism, organ, or tissue [6, 11, 12]. Bacterial cells containing gene(s) localized on plasmids transfer the gene(s) of interest to the new host cells. Delivery of genetic material is attained through entry of the entire bacterium into the target cells. Spontaneous or induced bacterial lysis leads to the release of the plasmid for future eukaryotic cell expression. Bactofection is a potent mechanism to express heterologous proteins into a large repertoire of cell types. Various bacterial species including *L. monocytogenes* and invasive *E. coli* have been examined as bactofection vectors [5].

This chapter describes the procedures for studying bacterial gene delivery in vitro and in vivo. It describes in detail how bactofection can be used for both in vitro and in vivo gene delivery systems. Also, this chapter will illustrate the growth conditions for bacteria and how they are prepared for systemic infection into tumors and the correct methods for injecting into tumor-bearing mice.

2   Materials

2.1   Tissue Culture Medium

1. Dulbecco's Modified Essential Medium (DMEM): Supplemented with 10 % (v/v) fetal calf serum (FCS) and 300 μg/ml L-glutamine.

2.2 Broth and Agar Solutions

All broth and agar solutions were prepared using distilled water according to the manufacturer's instructions. All broths were stored at room temperature.

1. Brain heart infusion (BHI): Dissolve 37 g in 1 l of distilled water. Mix well and distribute into final containers. Sterilize by autoclaving at 121 °C for 15 min.

2. Reinforced clostridial medium (RCM): Suspend 38 g in 1 l of distilled water. Bring to the boil to dissolve completely. Sterilize by autoclaving at 121 °C for 15 min.

3. De Man, Rogosa, Sharpe (MRS): Add 52 g to 1 l of distilled water at approximately 60 °C. Mix until completely dissolved. Dispense into final containers and sterilize by autoclaving at 121 °C for 15 min.

4. Luria-Bertani (LB): Add 20 g to 1 l of distilled water and mix until completely dissolved. Sterilize by autoclaving at 121 °C for 15 min.

5. Agar: Add 15 g of technical agar to 1 l of broth. Mix well and sterilize by autoclaving at 121 °C for 15 min. Dispense into Petri dishes and store at 4 °C till needed.

2.3 Antibiotics

1. Erythromycin: Dissolve 300 mg in 10 ml of 95 % ethanol for stock solution of 30 mg/ml. Sterilize by membrane filtration through 0.2 mm membrane. Store at –20 °C until needed.

2. Chloramphenicol: Dissolve 500 mg in 10 ml 0f 95 % ethanol for stock solution of 50 mg/ml. Sterilize by membrane filtration through 0.2 mm membrane. Store at –20 °C until needed.

3. Ampicillin: Dissolve 500 mg in 10 ml deionized water for stock solution of stock solution of 50 mg/ml. Sterilize by membrane filtration through 0.2 mm membrane. Store at –20 °C until needed.

3 Methods

3.1 Animal Conditions and Tumor Induction

1. Mice (Harlan Laboratories, Harlan, UK) were kept at a constant room temperature (22 °C) with a natural day/night light cycle in a conventional animal colony.

2. Standard laboratory food and water were provided ad libitum. Before experiments, the mice were afforded an adaptation period of at least 7 days.

3. Mice in good condition, without fungal or other infections, weighing 16–22 g, and of 6–8 weeks of age, were included in experiments.

4. Stocks of tumor cells were taken from liquid nitrogen and washed in DMEM media containing 10 % FCS.

5. These cells were inoculated into a tissue culture flask and grown till 80–90 % confluency under normal cell culture conditions (37°C, 5 % CO_2).

6. For tumor induction, the cells were washed twice in serum-free DMEM ($1,000 \times g$ for 5 min), and the minimum tumorigenic dose of cells was resuspended in final concentration of 200 µl of serum-free DMEM.

7. Prior to injection, the cells were mixed to ensure equal distribution of cells.

8. Using a 26 G needle, the cells were injected subcutaneously into the flank of 6–8-week-old mice.

9. Tumors were monitored mostly by alternate day measurements in two dimensions using a Vernier caliper.

10. Tumor volume was calculated according to the formula $V\frac{1}{4}ab2P/6$, where a is the longest diameter of the tumor, and b is the longest diameter perpendicular to diameter a.

3.2 Escherichia coli Growth Conditions and Injection into Mice

1. *E. coli* was taken from –80 °C stocks and streaked onto LB agar and left to grow overnight at 37 °C. If the strain contains a plasmid, an antibiotic will be present in the agar (chloramphenicol 50 µg/ml, erythromycin 250 µg/ml).

2. A single colony is then picked and inoculated into LB broth and placed at 37 °C where it is left shaking at 200 RPM overnight. Antibiotics are added if necessary.

3. A 1 % dilution is inoculated into fresh LB (~OD_{600} 0.05) and left to grow till it reaches early log phase (OD_{600} 0.6).

4. Once it has reached the correct OD_{600}, 1 ml of culture is spun at $13,500 \times g$ for 1 min, and the supernatant is removed.

5. The pellet is then resuspended in 1 ml PBS and spun once again at $13,500 \times g$ for 1 min. This is repeated twice more.

6. This corresponds to approximately 10^9 cfu/ml; this is then diluted down to 10^7 cfu/ml. 100 µl of this is then injected into the mouse tail vein giving a final concentration injected into the mice at 10^6 cfu/100 µl.

3.3 Bifidobacterium breve Growth Conditions and Injection into Mice

1. *B. breve* was taken from –80 °C stocks and inoculated into RCM broth and placed in an anaerobic hood at 37 °C and left to grow for 48 h. If the strain contains a plasmid, an antibiotic will be present in the broth (chloramphenicol 4 µg/ml).

2. A 1 % dilution is then inoculated into MRS broth containing 0.05 % cysteine and left in the anaerobic hood at 37 °C for 24 h. Antibiotics are added if required at the stated concentration.

3. A 1 % dilution is then inoculated into 100 ml MRS broth and left to grow anaerobically at 37 °C for overnight.

4. A 10 ml aliquot is then spun at $6,500 \times g$ for 5 min, and the supernatant is removed.

5. The pellet is washed three times in PBS containing 0.05 % cysteine–HCl, each time the sample is spun at $6,500 \times g$ for 5 min.

6. A 1 ml aliquot corresponds to approximately 10^9 cfu/ml; this is then diluted down to 10^7 cfu/ml. 100 µl of this is then injected into the mouse tail vein giving a final concentration injected into the mice at 10^6 cfu/100 µl.

3.4 Listeria monocytogenes Growth Conditions and Injection into Mice

1. *L. monocytogenes* was taken from −80 °C stocks and streaked onto BHI agar and left to grow overnight at 37 °C. If the strain contains a plasmid, an antibiotic will be present in the agar (chloramphenicol 7.5 µg/ml, erythromycin 10 µg/ml).

2. A single colony is then picked and inoculated into BHI broth and placed at 37 °C where it is left shaking at 200 RPM overnight. Antibiotics are added if necessary.

3. A 1 % dilution is inoculated into fresh BHI ($\sim OD_{600}$ 0.05) and left to grow till it reaches early log phase (OD_{600} 0.8–1.0).

4. Once it has reached the correct OD_{600}, 1 ml of culture is spun at $13,500 \times g$ for 1 min, and the supernatant is removed.

5. The pellet is then resuspended in 1 ml PBS and spun once again at $13,500 \times g$ for 1 min. This is repeated twice more.

6. This corresponds to approximately 10^9 cfu/ml; this is then diluted down to 10^6 cfu/ml. 200 µl of this is then injected into the mouse tail vein giving a final concentration injected into the mice at 10^5 cfu/100 µl.

3.5 Tail Vein Injections

1. For tail vein injection, the mice were anesthetized using 200 mg of xylazine and 2 mg of ketamine dissolved in PBS.

2. Once anesthetized, the tail of the mice were placed in lukewarm water to allow better visualization of the tail vein.

3. Using a 29 G needle, 100 µl–200 µl of the bacteria of interest is injected into the tail vein.

3.6 In Vivo Bioluminescent Imaging (BLI) of Bacteria in Tumors

1. In vivo BLI imaging was performed using the IVIS 100 (Caliper). At defined time points post bacterial administration, animals were anesthetized by intraperitoneal administration of 200 mg xylazine and 2 mg ketamine, and whole-body image analysis was performed in the IVIS 100 system for 2–5 min at high sensitivity.

2. Regions of interest were identified and quantified using Living Image software (Caliper) (Fig. 1).

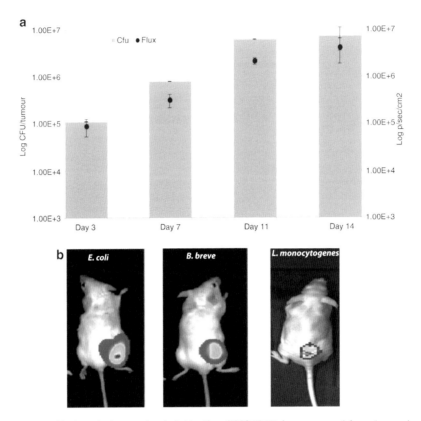

Fig. 1 (**a**) BLI results (Flux) and cfu counts of viable *E. coli* MG1655-*lux* recovered from tumor-bearing mice. Increase in bacterial numbers and plasmid gene expression in tumors was observed over time. (**b**) BLI of *E. coli* MG1655-*lux*, *B. breve* UCC 2003-lux, and *L. monocytogenes* EGDe-*lux* in tumor-bearing mice. Both *E. coli* and *L. monocytogenes* contain the p*16slux* plasmid, while *B. breve* was tagged with pLUXMC3 plasmid. All BLI were performed using IVIS 100 (Caliper)

3.7 In Vitro Bactofection Assay

1. Cell lines were maintained in DMEM + 10 % FCS, and no anti-biotics were added to cell lines during propagation. Cell lines were routinely cultured at 37 °C in 5 % CO_2.

2. Cell lines were seeded at until confluency in 24-well plates (Falcon), and *L. monocytogenes* was infected at MOI of 50:1.

3. Cell types were invaded for 1 h at 37 °C in 5 % CO_2, washed once with Dulbecco's PBS (Sigma), and then overlaid with DMEM containing 10 µg/ml gentamicin for 1 h.

4. Following gentamicin treatment, the medium was replaced with DMEM (10 % FCS) only or with ampicillin (20, 100 or 500 µg/ml). Ampicillin treatment was for 24 or 48 h and was refreshed after 12 h.

5. Monolayers were washed a further three times with PBS to remove residual antibiotic and then lysed with 1 ml of ice-cold sterile water. Bacterial cells were enumerated by serial dilution in PBS and plated on BHI agar.

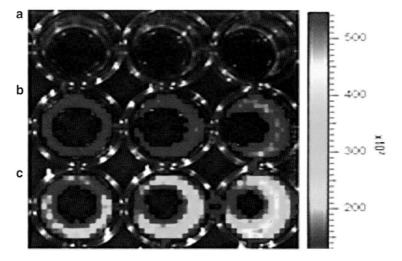

Fig. 2 In vitro bactofection assay containing *L. monocytogenes* pFXEmLuc plasmid to allow measurement of luciferase activity after treatment with ampicillin. (**a**) No treatment of cells with ampicillin, (**b**) FLuc expression 24 h post-ampicillin treatment of *L. monocytogenes* pFXEmLuc, (**b**) FLuc expression 48 h post-ampicillin treatment of *L. monocytogenes* pFXEmLuc. All luciferase activity was measured using IVIS 100 (Caliper)

6. To assess firefly luciferase expression, monolayers were washed three times with PBS and resuspended in 250 μl reporter lysis buffer (Promega, Madison, WI).

7. The samples were subjected to one freeze–thaw cycle to obtain complete lysis. Lysis solutions were centrifuged (1 min 13,000 × *g*), and the supernatants were transferred to a 24-well plate.

8. Levels of firefly luciferase protein expression were determined by adding an equal volume of the substrate luciferin. Luminescence was measured in relative light units (RLU) (in photons) in a IVIS100 (Perkin Elmer, MA, USA) (Fig. 2).

3.8 In Vivo Bactofection Gene Delivery

1. The mice were randomly divided into experimental groups and subjected to specific experimental protocols. For tumor experiments, the mice were treated at a tumor volume of approximately 100 mm^3 in volume (5–7 mm major diameter).

2. When applicable, tumors were injected with *Listeria monocytogenes* as follows: 100 μl *Listeria monocytogenes* suspension (10^6 cfu/ml) was administrated intratumorly by injection into the center of the tumor tissue using a 29 G needle.

3. For *Listeria monocytogenes* mediated transfection, after 6 days, if applicable, a single dose of 100 μl ampicillin (100 mg/ml) was administrated for 3 consecutive days.

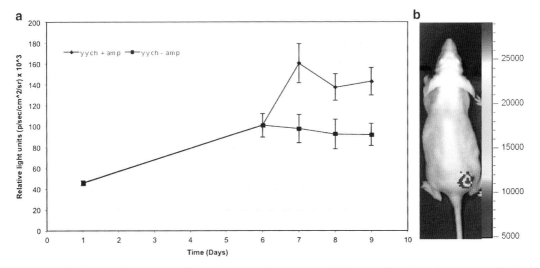

Fig. 3 In vivo bactofection assay utilising *L. monocytogenes* ampicillin sensitive strain for improved intra-tumoural gene delivery. (**a**) *L. monocytogenes* pFXEmLuc measurement of firefly luciferase (FLuc) activity in the presence and absence of ampicillin. Luminescence acts as a read-out for bactofection, since only eukaryotic cells express the FLuc construct. (**b**) BLI of in vivo bactofection utilising *L. monocytogenes* pFXEmLuc

4. Following euthanasia, assessment of cfu/ml in liver and spleen was performed by placing the organs in 5 ml of PBS. The organs were then broken down to a single cell suspension using a 70 μm nylon cell strainer (BD Falcon). In a similar manner, cfu/ml in tumor tissue was determined.

3.9 Whole-Body Imaging for In Vitro Bactofection Gene Delivery

1. In vivo luciferase activity from tissues was analyzed at set time points post-transfection as follows: 80 μl of 30 mg/ml firefly luciferin (Biosynth, Basel, Switzerland) was injected intraperitoneally or intratumorally.

2. Mice were anesthetized by intraperitoneal (i.p.) administration of 200 μg xylazine and 2 mg ketamine.

3. 10 min post-luciferin injection, live anesthetized mice were IVIS imaged for 3 min at high sensitivity. All data analysis was carried out on the Living Image software package (Perkin Elmer) (Fig. 3).

References

1. Cummins J, Tangney M (2013) Bacteria and tumours: causative agents or opportunistic inhabitants? Infect Agent Cancer 8(1):11. doi: 10.1186/1750-9378-8-11

2. Cronin M, Stanton RM, Francis KP, Tangney M (2012) Bacterial vectors for imaging and cancer gene therapy: a review. Cancer Gene Ther. doi:10.1038/cgt.2012.59

3. Hoption Cann SA, van Netten JP, van Netten C (2003) Dr William Coley and tumour regression: a place in history or in the future. Postgrad Med J 79:672–680

4. Morrissey D, O'Sullivan GC, Tangney M (2010) Tumour targeting with systemically administered bacteria. Curr Gene Ther 10 (1): 3–14. doi: ABS-17 [pii]

5. Baban CK, Cronin M, O'Hanlon D et al (2010) Bacteria as vectors for gene therapy of cancer. Bioeng Bugs 1:385–394

6. Tangney M (2010) Gene therapy for cancer: dairy bacteria as delivery vectors. Discov Med 10:195–200

7. Bermudes D, Low B, Pawelek J (2000) Tumor-targeted Salmonella. Highly selective delivery vectors. Adv Exp Med Biol 465:57–63

8. Sznol M, Lin SL, Bermudes D et al (2000) Use of preferentially replicating bacteria for the treatment of cancer. J Clin Invest 105:1027–1030

9. Cronin M, Akin AR, Collins SA et al (2012) High resolution in vivo bioluminescent imaging for the study of bacterial tumour targeting. PLoS One 7:e30940

10. Cronin M, Morrissey D, Rajendran S et al (2010) Orally administered bifidobacteria as vehicles for delivery of agents to systemic tumors. Mol Ther 18:1397–1407

11. van Pijkeren JP, Morrissey D, Monk IR et al (2010) A novel Listeria monocytogenes-based DNA delivery system for cancer gene therapy. Hum Gene Ther 21:405–416

12. Palffy R, Gardlík R, Hodosy J et al (2006) Bacteria in gene therapy: bactofection versus alternative gene therapy. Gene Ther 13:101–105

Chapter 14

Synthesis of a Smart Nanovehicle for Targeting Liver

Arnab De, Sushil Mishra, Seema Garg, and Subho Mozumdar

Abstract

A protocol for the synthesis of a smart drug delivery system based on gold nanoparticles has been described in this chapter. The synthesized drug delivery system has been shown to release the bioactive material in response to an intracellular stimulus (glutathione concentration gradient) and hence shown to behave in an intelligent manner. Gold nanoparticles have been employed as the core material with the surface functionalities of thiolated PEG. PEG owing to its non-immunogenicity and non-antigenicity would impart considerable stability and longer in vivo circulation time to the gold nanoparticles. The end groups of PEG chains have been derivatized with functional groups like aldehyde (–CHO) and amine (–NH$_2$) which could behave as flexible arms for the attachment of "target specific ligands" and other bioactive substances. Lactose, a liver targeting ligand, has been employed as the target specific moiety. A Coumarin derivative has been synthesized and used as the model fluorescent tag as well as a linker to examine the glutathione-mediated release through fluorescence spectroscopy and for the conjugation of bioactive molecules, respectively. A check for the cytocompatibility of the synthesized nanovehicle on the cultured mammalian cell lines has also been carried out. Finally, in the latter parts of the chapter (mimicking the in vivo conditions), time-dependent in vitro release of the model fluorescent moiety has also been analyzed at different glutathione concentrations.

Key words Smart drug delivery system, Targeting liver, Gold nanoparticles, Thiolated PEG, Lactose, Coumarin, Glutathione, Fluorescence spectroscopy

1 Introduction

Nanotechnology is expected to revolutionize the field of biomedical sciences. Various systems at the scale of nanometer have been developed for the effective delivery of drug and other bioactive substances. These nanostructures have been postulated to play important role in sensing, imaging, and therapy of diseases [1]. Though the number of such nanostructures are increasing exponentially, targeting drugs to particular tissue and then releasing the payload in the communicative response to the specific stimuli

---

Arnab De, Sushil Mishra, and Seema Garg have contributed equally to this work.

Kewal K. Jain (ed.), *Drug Delivery System*, Methods in Molecular Biology, vol. 1141,
DOI 10.1007/978-1-4939-0363-4_14, © Springer Science+Business Media New York 2014

remains a major challenge. In order to overcome the challenge, researchers all around the world have come up with some Smart Drug Delivery Systems (SDDS). Smart delivery systems are self-sufficient to regulate their activities such as "on and off" of payload release in response to a specific stimuli. Several release approaches have been subsequently developed, relying either on external stimulus such as light [2, 3], electric field [4–6], ultrasound [7, 8], etc. or on internal stimulus such as pH [9–12], glucose [13–16], enzymes [17–19], antigen [20–22], redox/thiol [23, 24], temperature [25–28], etc.

Glutathione (tripeptide GSH; most abundant thiol inside the cells) is one such intracellular stimulus which is present inside the biological systems and can offer a triggered intracellular release of payload due to its concentration gradient across the cells [29]. The concentration of glutathione in the blood plasma is 10 μM [30], whereas the intracellular concentration varies from 1 to 10 mM (the maximum being in hepatic cells) [31]. This 1,000-fold significant difference in the intracellular verses extracellular thiol level makes GSH an effective trigger for selective intracellular release through thiol place exchange reaction [32]. At the site of action, GSH can displace the existing thiol through the reduction of the linkage that binds the thiol on to the carrier (viz., nanoparticles) surface, rendering the immediate release of the bioactive molecule in the physiological environment. Previous work has demonstrated the use of GSH as an effective stimulus for the controlled delivery of DNA [33, 34] and other drug moieties from cross-linked polymeric shells [35–37].

Since thiol place exchange reactions forms the basis of the GSH activity as an intracellular drug release stimulus, a drug carrier with strong affinity towards thiols becomes a prerequisite before exploiting GSH for the development of Smart Drug Delivery Systems. Metallic nanoparticles especially Gold nanoparticles as drug carrier holds the prime position in this context. Gold nanoparticles have been known since long to serve as inert and nontoxic vehicles for drug delivery [38, 39], biosensing [40], diagnostics [41], imaging [42, 43], and gene delivery applications [44–46]. They can be easily synthesized with core sizes ranging from 1 to 150 nm [47].

Bare gold nanoparticles, however, pose some complications that need to be addressed before their biomedical use can be envisioned. A few of such complications are instability due to the absorption of protein and other molecules on their highly chemical reactive surface and nonspecific uptake of the particles (size-dependent) by the reticular–endothelial system [48]. In the recent times, these challenges have been overcome by coating the surface of gold nanoparticles with a layer of hydrophilic and biocompatible polymer such as polyethylenimine [49] and poly(ethylene glycol) [50, 51], etc. Such types of coatings prevent the aggregation of gold nanoparticles in high ionic strength environment and support long-circulation of nanoparticles in the in vivo systems that are desired for passive targeting to tumor and inflammatory sites [46]. Moreover, conjugating gold nanoparticles with some specific functional moiety can also direct them for targeting specific organs or tissues [54, 55].

Hence, the importance of both gold nanoparticles as the drug carrier and glutathione as the stimulus can be utilized for providing an effective and selective means of controlled intracellular release. Therefore in the present work an application of gold–thiol chemistry to design hepatic cell-specific nanodrug delivery vehicle featuring gold core with surface functionalities of thiolated polyethylene glycol (PEG) has been shown. The PEG moieties are further conjugated with the liver targeting ligand (lactose) and a fluorescent molecule (7-aminocoumarin-3-carboxylic acid) for various biomedical applications. The fluorescent moiety can serve both as a linker for bioconjugation with molecules of biological importance as well as a model to quantify the payload release in response to variable glutathione concentrations through quantitative fluorescence measurements. Cytotoxicity assessment of the synthesized nanovehicle is carried out on the cultured human cell lines (human alveolar basal epithelial, A549) employing the MTT cell viability test [52, 53].

2   Materials

1. Gold chloride ($HAuCl_4\cdot3H_2O$, 49–50 % Au) is procured from Thomas Baker (Mumbai, India). An aqueous solution of 0.05 M gold chloride is freshly prepared before use.

2. A stock surfactant solution of nonionic surfactant (0.1 M) is prepared by dissolving weighed amount of surfactant in cyclohexane and clearing the appeared turbidity in the solution by slowly adding double-distilled water under continuous stirring.

3. Methoxy PEG 2000 (polyethylene glycol) and methoxy PEG 5000 are obtained from Sigma-Aldrich. Both PEG products are azeotropically refluxed with toluene to remove the excess water prior to their usage.

4. Salicylaldehyde, malonic acid, tosyl chloride, lithium aluminum hydride, potassium thioacetate, sodium azide, reduced glutathione, lactose, Amberlite resin (IR-120B), iodine granules, and tin chloride are products of Aldrich Chemicals.

5. Acids are obtained from Thermo Scientific and are used without any further purification.

6. Dulbecco's Modified Eagle's Medium is obtained from Sigma-Aldrich and stored in dark at 4 °C.

7. 10 % fetal calf serum is procured from Invitrogen and stored at 4 °C.

8. 1 % antibiotic (streptomycin) is obtained from Sigma-Aldrich and stored at 4 °C.

9. 96-well plates are supplied from Nunc, Thermo Scientific.

10. Thiazolylbluetetrazoliumblue(MTT)[(3-(4,5-Dimethylthiazol-2-yl)-2,5-Diphenyltetrazolium Bromide)] is procured for Sigma-Aldrich. The stock working solution contains 0.5 mg/mL of MTT and stored at 4 °C.

11. Dimethylsulfoxide (DMSO) and phosphate buffer saline powder (pH 7.4) are supplied from Sigma-Aldrich.

12. Double-distilled water prepared freshly in laboratory is used throughout the experiment.

3   Methods

3.1   Synthesis of Gold Nanoparticles

Reverse microemulsion technique is employed to synthesize gold nanoparticles by using a biocompatible non-ionic surfactant as the amphiphile, cyclohexane as the continuous phase and aqueous salt solutions as the dispersed phase. The steps involved in the typical synthesis procedure are listed below:

1. Prepare microemulsion A by adding 180 μL of 0.05 M gold chloride to 25 mL of stock surfactant solution under constant magnetic stirring.

2. Prepare microemulsion B by adding 180 μL of 0.25 M sodium borohydride to a separate 25 mL of stock surfactant solution under constant magnetic stirring.

3. Equilibrate both microemulsions (A and B) for 45 min through vigorous magnetic stirring.

4. Add microemulsion B to microemulsion A so as to form a homogeneous microemulsion mixture (microemulsion C). Leave Microemulsion C on stirring for 4–5 h.

5. Observe Microemulsion C for the hydrodynamic radii of the gold nanoparticles on a Dynamic Light Scattering (DLS) instrument (*see* **Note 1**) and then store it in the sealed state (*see* **Note 2**) at temperature ranging from 30 to 50 °C.

3.2   Synthesis of Targeting Moiety Lactonolactone

Lactonolactone is synthesized by oxidizing the sugar moiety Lactose [54].

1. Dissolve 3.7 g of iodine granules in 40 mL of methanol at 40 °C and add 2 g of lactose (dissolved in minimum amount of hot water) to it. The reaction mixture is stirred for 2 h.

2. Add dropwise a concentrated solution of potassium hydroxide (40 mg/mL) in methanol to the reaction mixture kept at 40 °C until the color of iodine disappeared.

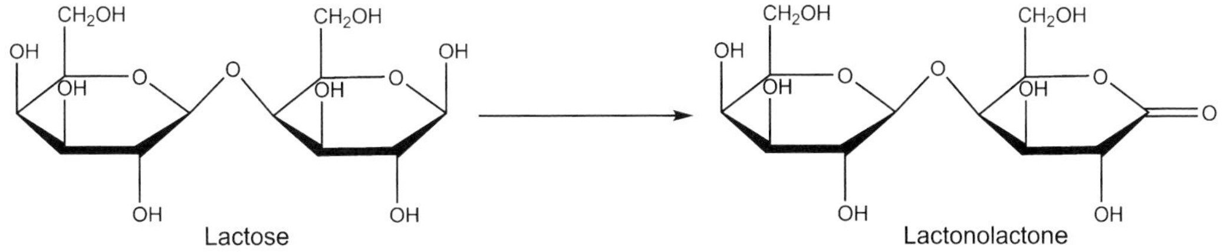

Fig. 1 Synthesis of lactonolactone from lactose

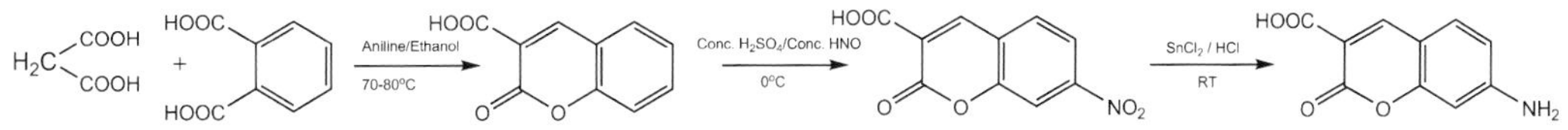

Fig. 2 Synthesis of 7-aminocoumarin 3-carboxylic acid

3. Cool the solution externally by keeping it in an ice bath which leads to the precipitation of a crystalline product.

4. Filter the product and wash it repeatedly with cold methanol. Recrystallize the product by using water–methanol system.

5. Potassium salt of lactobionic acid is formed which is further converted to its free acid form by passing it through a column of acidic Amberlite resin.

6. The aqueous elute is freeze-dried (*see* **Note 3**) and evaporated several times with methanol to get the final lactonolactone product (highly viscous colorless oil) (*see* Fig. 1).

3.3 Synthesis of Fluorescent Moiety 7-Aminocoumarin 3-Carboxylic Acid

The synthesis of the fluorescent moiety is performed in three consecutive steps (*see* Fig. 2).

3.3.1 Synthesis of 3-Carboxycoumarin

The first step involves the synthesis of 3-carboxycoumarin from salicylaldehyde and malonic acid [55]. The steps involved in the synthesis are as follows:

1. Reflux the mixture of salicylaldehyde (6 mmol), malonic acid (6 mmol), and 0.1 mL of aniline in 20 mL of ethanol for 3 h.

2. Evaporate the solvent (ethanol) in vacuum (*see* **Note 4**) and a yellow-gray mass is obtained.

3. Add 50 mL of double-distilled water to the yellow-gray mass and stir it for 15–20 min followed by filtration. Repeat the same procedure for three to four times.

4. Take the filtrate in 50 mL of double-distilled water and add 1 mL of conc. HCl to it.

5. Heat the emulsion so formed for 2 h at 70 °C.

6. Precipitate obtained is filtered, washed, and air-dried (Product A).

3.3.2 Nitration of 3-Carboxycoumarin

Next step involves controlled nitration of 3-carboxycoumarin (Product A) at cold temperatures to synthesize 7-nitrocoumarin 3-carboxylic acid.

1. Take 200 mg of 3-carboxy coumarin (Product A) in an ice cold jar.

2. Add 2 mL of an ice-cold mixture of concentrated sulfuric acid and nitric acid (1:1) to it in a dropwise manner at the maintained temperature of 0 °C taking a time span of about 30 min.

3. Stir the reaction mixture for about 5–10 min at 0–4 °C.

4. After the completion of the reaction (*see* **Note 5**), add double-distilled water (10–15 mL) to cease the nitration process.

5. Filter the precipitated product and dry it under vacuum conditions (*see* **Note 6**).

6. Recrystallize the crude product with methanol to get crystals of 7-nitro-3-carboxycoumarin (Product B).

3.3.3 Reduction of 7-Nitro-3-Carboxycoumarin

Finally, Product B is reduced with $SnCl_2$–HCl mixture to yield the final product as a bright yellow solid [56].

1. Dissolve tin chloride (2 mmol) in minimum amount of concentrated HCl at 60 °C.

2. Add 7-nitrocoumarin 3-carboxylic acid (Product B; 1 mmol) to it and reflux the reaction mixture for about 6 h under same condition of temperature.

3. Leave the reaction mixture on stirring overnight at room temperature.

4. Following the completion of reaction (*see* **Note 7**), neutralize the reaction mixture with aqueous solution of sodium bicarbonate till the pH of 6.8 is achieved (*see* **Note 8**).

5. Treat the reaction mixture repeatedly with dry diethyl ether.

6. Filter the precipitated final product and dry it under vacuum conditions (*see* **Note 6**).

3.4 Synthesis of SH-PEG-NH₂ from Monomethoxy PEG 5000

The synthesis of hetero bifunctional PEG derivative from monomethoxy PEG 5000 involves a multiple step synthesis (*see* Fig. 3). The synthetic procedures have been referred from a few publications [1, 57] and are performed with some modifications.

3.4.1 Derivative AcS-PEG-OH (3.i)

1. Dissolve monomethoxy PEG 5000 (1.0 eq) in a minimum amount of DMF and add potassium thioacetate (10.0 eq) under continuous stirring and inert atmosphere for about 24 h (*see* **Note 9**).

2. After the completion of the reaction (*see* **Note 10**), the crude reaction mixture is treated with dichloromethane (CH_2Cl_2).

3. Wash excess of thioacetate by adding equal parts of saturated solution of ammonium chloride (NH_4Cl) and brine solution.

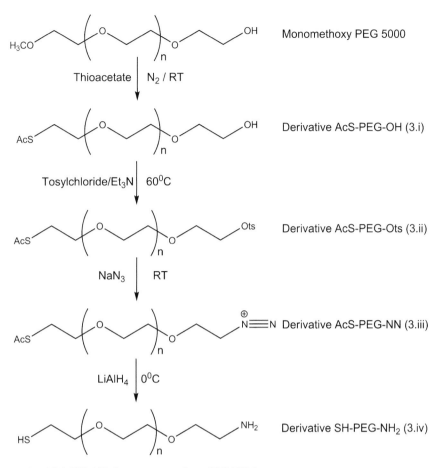

Fig. 3 Synthesis of SH-PEG-NH$_2$ from monomethoxy PEG 5000

4. Perform the liquid–liquid extraction (*see* **Note 11**) and then separate aqueous and organic layer followed by four to five times extraction of the aqueous layer with dichloromethane (CH$_2$Cl$_2$).

5. Finally, the organic layer aliquots are combined, evaporated, and purified over alumina (methanol–CH$_2$Cl$_2$ system) to get the final crude product (yellow oil with foul smell).

6. The yellow oil is treated with diethyl ether to give a pale yellow solid as the final product (*see* **Note 12**).

3.4.2 Derivative AcS-PEG-Ots (3.ii)

Thioacetate derivative of P3EG was tosylated using the following procedure:

1. Dissolve AcS-PEG-OH (1.0 eq) in minimum amount of toluene and add triethylamine base (3.0 eq) and *p*-tosylchloride (1.5 eq) to it.

2. Initially stir the reaction mixture for 5 h at temperature of 60 °C and further it is stirred for another 10 h at ambient (25–30 °C) temperature.

3. After the completion of the reaction (*see* **Note 10**), remove the solvent from the reaction mixture by using rotary evaporator (*see* **Note 4**).

4. Dissolve the crude product in dichloromethane (CH_2Cl_2) and add 30 mL of 0.25 M aqueous HBr solution and 10 mL brine solution for washing the excess of triethylamine and tosylchloride, respectively.

5. Perform the liquid–liquid extraction (*see* **Note 11**) and then separate aqueous and organic layer followed by four to five times extraction of the aqueous layer with dichloromethane (CH_2Cl_2).

6. Finally, the organic layer aliquots are combined, evaporated, and purified over alumina (methanol–CH_2Cl_2 system) to get the product (pale yellow colored oil), which upon trituration with diethyl ether yields a pale yellow colored final product (*see* **Note 12**).

3.4.3 Derivative AcS-PEG-N≡N (3.iii)

1. Dissolve derivative AcS-PEG-Ots (3.ii) (1.0 eq) in minimum amount of DMF and add sodium azide (1.25 eq) under inert atmosphere at ambient conditions (*see* **Note 9**).

2. Stir the reaction mixture at the same conditions for 24 h.

3. The formation of product was confirmed through TLC (CH_2Cl_2–MeOH) (*see* **Note 10**).

4. Add dry diethyl ether to precipitate out the crude product from the reaction mixture.

5. Purify by column chromatography over alumina (EtOAc–MeOH) to yield the final product (pale yellow oil) which upon trituration with ether gives a pale yellow colored solid (*see* **Note 12**).

3.4.4 Derivative SH-PEG-NH₂ (3.iv)

1. Maintain the solution of $LiAlH_4$ (5.0 eq) in dry DMF at –10 to 0 °C under inert atmosphere (*see* **Note 9**) and add the solution of AcS-PEG-N≡N (1.0 eq) dissolved in DMF into it in a drop wise manner.

2. Stir the reaction mixture at the maintained conditions of temperature and inertness for 4 h.

3. Perform the Ellman's test for thiol group and ninhydrin test for the amino group to analyze the progress of the reaction (*see* **Notes 13** and **14**).

4. After the completion of the reaction add double-distilled water very cautiously to precipitate lithium hydroxide and then filter it over a pad of celite followed by washing it repeatedly with ethanol.

5. Concentrate the filtrate and dissolved it in a minimum amount of CH_2Cl_2 and purify over alumina (EtOAc–MeOH system) to yield colorless oil. Treating it with dry diethyl ether yields a final product of an off-white color solid (*see* **Notes 15** and **17**).

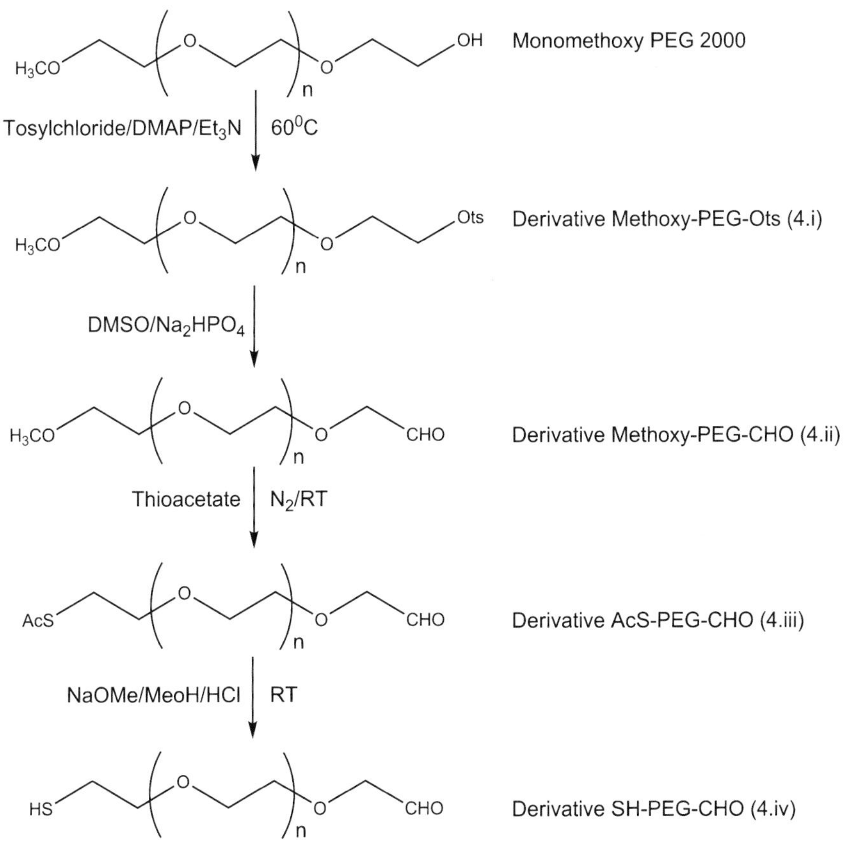

Fig. 4 Synthesis of SH-PEG-CHO from monomethoxy PEG 2000

3.5 Synthesis of SH-PEG-CHO from Monomethoxy PEG 2000

The hetero-bifunctional polyethylene glycol (PEG) derivative containing active end-groups thiol and aldehyde is prepared from monomethoxy PEG (*see* Fig. 4). The synthetic procedure is performed with some modifications of that described in publications [1, 57, 58].

3.5.1 Derivative Methoxy-PEG-Ots (4.i)

1. Dissolve purified monomethoxy PEG (1.0 eq) in minimum amount of toluene and add Et₃N (3.0 eq), DMAP (catalytic amount; 0.25 eq), and *p*-tosylchloride (1.5 eq) to it.

2. Heat the reaction mixture in an oil bath maintained at 80 °C for 72 h.

3. After completion of the reaction (*see* **Note 10**) cool the reaction mixture to room temperature and remove the solvent from the reaction mixture by using rotary evaporator (*see* **Note 4**).

4. The crude oil product so obtained is treated with CH₂Cl₂.

5. Excessive reagents are removed through vigorous washings with saturated solution of NaHCO₃, brine and 0.25 M aqueous HBr.

6. Perform the solvent extraction (*see* **Note 11**) and separate the aqueous and organic layers. Collect the organic layer and repeat the extraction process four to five times.

7. The organic layer aliquots are combined, evaporated, and purified over alumina (CH_2Cl_2–MeOH system) to obtain final product (colorless oil).

8. Trituration with dry diethyl ether gives the final product as off-white colored solid (*see* **Note 12**).

3.5.2 Derivative Methoxy-PEG-CHO (4.ii)

1. Take the solution of MeO-PEG-Ots (1.0 eq) in 15 mL DMSO and treat it with Na_2HPO_4 (20.0 eq). Stir the mixture in an oil bath maintained at 100 °C for 20 h.

2. After the completion of the reaction (*see* **Note 10**) cool the reaction mixture and filter. The filtrate is precipitated out several times with dry diethyl ether which yielded semi-solid off-white colored precipitate.

3. Dissolve the precipitate in minimum amount of water to perform dialysis (*see* **Note 16**) against double-distilled water followed by lyophilization (*see* **Note 3**) to obtain the final product as an off-white colored solid (*see* **Note 15**).

3.5.3 Derivative AcS-PEG-CHO (4.iii)

1. Add potassium thioacetate (1.5 eq) to a solution of derivative Methoxy-PEG-CHO (4.ii) (1.0 eq) dissolved in minimum amount of DMF under complete inert atmosphere (*see* **Note 9**).

2. Stir the reaction mixture continuously at room temperature for 24 h maintaining the inertness of the system.

3. After the completion of the reaction (*see* **Note 10**), the crude reaction mixture is treated with dichloromethane (CH_2Cl_2).

4. Wash excess of thioacetate by adding equal parts of saturated solution of ammonium chloride (NH_4Cl) and brine solution.

5. Perform the liquid–liquid extraction (*see* **Note 11**) and then separate aqueous and organic layer followed by four to five times extraction of the aqueous layer with dichloromethane (CH_2Cl_2).

6. Finally, the organic layer aliquots are combined, evaporated, and purified over alumina (methanol–CH_2Cl_2 system) to get the final crude product.

7. The crude product is treated with diethyl ether to give a pale yellow solid as the final thioacetated product (*see* **Note 15**).

3.5.4 Derivative SH-PEG-CHO (4.iv)

1. Add derivative AcS-PEG-CHO (4.iii) (1.0 eq) dissolved in minimum amount of degassed methanol to a solution of sodium methoxide (NaOMe; 5.0 eq) in MeOH.

2. Stir the reaction mixture overnight at room temperature.

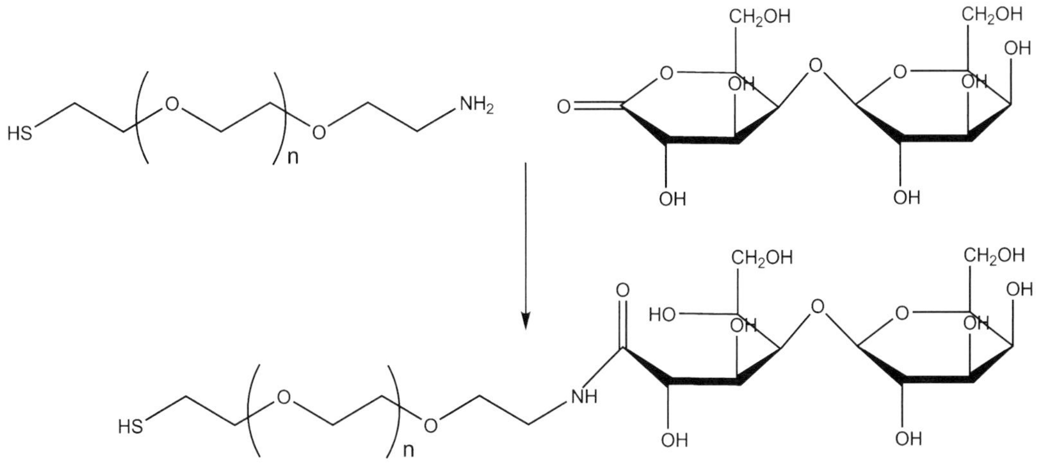

Fig. 5 Coupling of lactonolactone to SH-PEG-NH$_2$ (Adduct 1)

3. Acidify the mixture to pH 1–2 using 0.1 N HCl.

4. Evaporate the solvent from the reaction mixture over rotary evaporator (*see* **Note 4**) to yield the crude product.

5. Purify the crude product by silica gel chromatography (CH$_2$Cl$_2$–MeOH system) to yield the bifunctional derivative as colorless oil, which upon trituration with dry diethyl ether give off-white colored solid (*see* **Notes 15** and **17**).

3.6 Coupling of Lactonolactone to SH-PEG-NH$_2$ [1]

1. Add lactonolactone (8.0 eq) and SH-PEG-NH$_2$ (1.0 eq) in minimum amount of DMF and carry out coupling reaction under inert atmosphere (*see* **Note 9**) at 100 °C for 30 h.

2. After the completion of the reaction (*see* **Note 10**), the final product is precipitated out with MeOH–dry ether (1:1) mixture and finally with dry diethyl ether. The product is dried under vacuum (*see* **Note 15**) overnight to give brown colored solid (adduct 1) (*see* Fig. 5).

3.7 Coupling of 7-Aminocoumarin 3-Carboxylic Acid to SH-PEG-CHO

1. Dissolve coumarin derivative (1.0 eq) in minimum amount of methanol and adjust pH to 8.0 with triethylamine.

2. Add 4.0 eq of aldehyde (4.iv) and 25 μL of 5 M sodium cyanoborohydride to the stirring solution of coumarin under inert atmosphere (*see* **Note 9**).

3. The reaction mixture is stirred overnight at room temperature maintaining the inertness of the system.

4. After the completion of the reaction (*see* **Note 10**), the solvent is evaporated and the solid residue is treated with CH$_2$Cl$_2$. Filter the solution to remove the unreacted excessive reagents.

5. Purify the filtrate by silica chromatography (CH$_2$Cl$_2$–MeOH) to yield the final product as a bright yellow-colored solid (adduct 2) (*see* **Note 15** and Fig. 6).

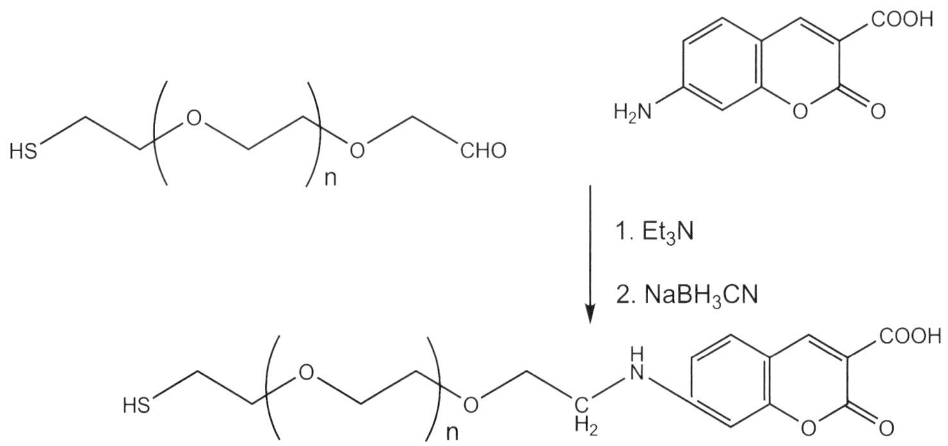

Fig. 6 Coupling of Coumarin derivative (7-Aminocoumarin 3-Carboxylic Acid) to SH-PEG-CHO (Adduct 2)

3.8 Conjugation of Adduct 1 and Adduct 2 with Gold Nanoparticles

1. Treat the surfactant-protected gold nanoparticles with benzene–methanol (3:1) mixture in order to disrupt the surfactant assemblies before coupling reaction.

2. Extract out the bare gold nanoparticles (*see* **Note 18**), and then wash and dispersed them in 5 mL of phosphate buffer at pH 7.4.

3. Disperse separately, adduct 1 and adduct 2 in 2 mL of the PBS buffer at pH 7.4

4. Maintain the inert condition (*see* **Note 9**) for reaction and add 500 μL of adducts 1 and 2 to 5 mL of the gold colloid solution.

5. Incubate the reaction mixture at room temperature for 24 h maintaining the inertness of the system. The colloidal suspension is then centrifuged ($11,000–16,000 \times g$) to separate the functionalized nanoparticles.

6. Wash extensively the nanoparticles with ethanol.

7. Finally, the washed nanoparticles are re-dispersed in 5 mL of the phosphate buffer of pH 7.4 (*see* Fig. 7).

8. Measure the size of nanoparticle by using DLS (*see* **Note 1**) and TEM analysis (*see* **Note 19** and Fig. 8)

3.9 Glutathione-Mediated In Vitro Release Studies

Glutathione-mediated release of the fluorescent moiety (7-amino 3-carboxy coumarin) from the surface of pEGylated gold nanoparticles is quantified as a function of variable glutathione concentration and time through fluorescence spectroscopy (*see* Fig. 9).

3.9.1 Concentration-Dependent Release Studies

1. Prepare various concentrations of reduced glutathione ranging from 2.5 μM to 10 mM in PBS buffer and measure their UV absorption (*see* **Note 20**) and fluorescence values (*see* **Note 21**).

2. Take 3 mL aliquot of the glutathione solution at all the prepared concentrations in separate sample vials.

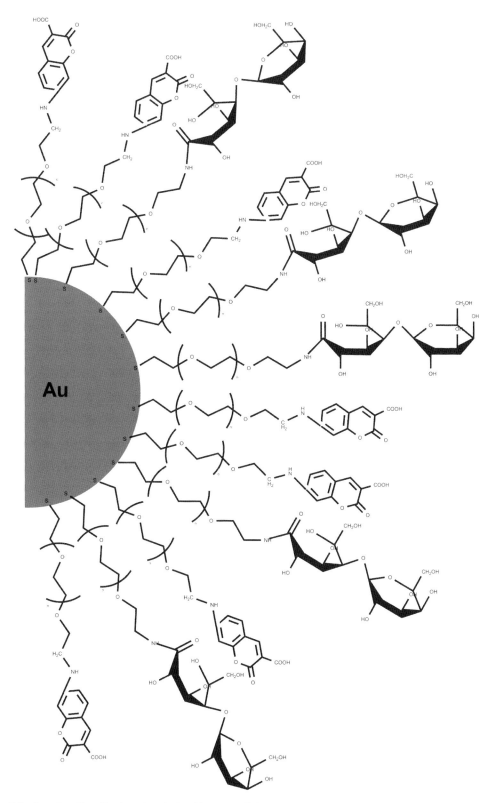

Fig. 7 Surface functionalized conjugated gold nanoparticle

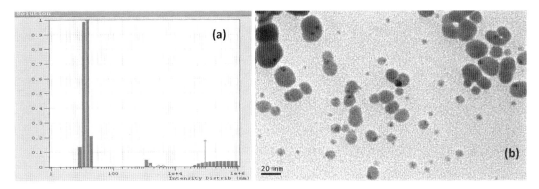

Fig. 8 DLS data (**a**) and TEM Micrograph (**b**) of conjugated gold nanoparticles

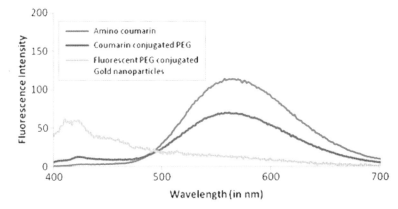

Fig. 9 Fluorescence spectra showing quenching of fluorescence by gold nanoparticles. The fluorescence spectrum of coumarin shows the characteristic peak at λ_{max} of 572 nm with a I_{max} of 113. The coumarin conjugated PEG shows peak at the same λ_{max} i.e., 572 nm but with lower intensity ($I_{max}=69$). On the other hand, the fluorescence of the coumarin (covalently attached to PEG 2000 derivatives) is quenched in the conjugated gold nanoparticles indicating that the latter behave as efficient fluorescence quenchers

3. Maintaining the temperature at 37 °C, add 500 µL of bioconjugated gold colloid in PBS buffer to all the glutathione aliquots.

4. Incubate the resulting colloids for 6 h at 37 °C

5. After 6 h, quantify the release of fluorescent moiety from the gold surface by measuring the fluorescence of colloids (*see* **Note 21**).

6. Perform a blank study by incubating 500 µL of gold colloid with 3 mL of PBS buffer at 37 °C for 6 h and then measure its fluorescence (*see* Fig. 10).

3.9.2 Time-Dependent Release Studies

1. Maintain the temperature at 37 °C, add 500 µL of bioconjugated gold colloid in PBS buffer to 3 mL of glutathione aliquots at concentration of 10 µM and 10 mM.

2. Incubate the colloid at 37 °C and measure the fluorescence (*see* **Note 21**) at different time point up till 24 h.

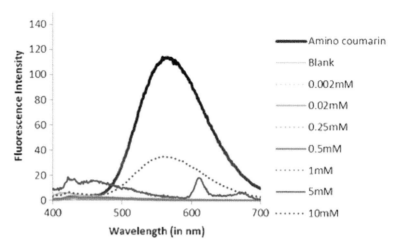

Fig. 10 Shows the release studies for the functionalized gold nanoparticles incubated with different concentration of glutathione (starting from 2.5 to 10 mM) at physiological temperature for a time span of 6 h. The fluorescence moiety remains quenched in the presence of glutathione concentration of 1 mM or lower; whereas, at and above 5 mM, the peak appears to be significant. At low concentrations of glutathione (≤ 1.0 mM), the biological thiol is not sufficient to displace the thiols on the nanoparticle surface or the thiols on the nanoparticle surface are in more concentration than glutathione thereby shifting the equilibrium towards conjugation of nanoparticles with thiolated PEG. But as soon as the concentration increases to 10 mM, the glutathione act as trigger to displace thiolated PEG from the nanoparticle surface. The thiolated PEG disassembled from the nanoparticle surface and as the consequence the influence of gold nanoparticles on the fluorescence characteristics of coumarin derivative disappeared resulting in a sharp signal peak in its respective fluorescence spectrum

3. Perform a blank study by incubating 500 μL of gold colloid with 3 mL of PBS buffer at 37 °C for 24 h and measuring its fluorescence at different time points (*see* Fig. 11).

3.10 In Vitro Cytotoxicity Assessment of Conjugated Gold Nanoparticles

The cytocompatibility studies of the surface functionalized and bare gold nanoparticles is performed on A549 human cell lines via MTT cell viability assay. Human alveolar basal epithelial (A549) cells are cultured in Dulbecco's modified Eagle's medium (DMEM) supplemented with 10 % fetal bovine serum and 1 % antibiotic. The steps involved in the cytotoxicity assessment are as follows.

1. Maintain the cells at 37 °C in a 5 % CO_2 incubator.

2. Plate the cells (approximately, 1×10^5 cells) in 96-well plates.

3. Add separately 1 mg of bare and conjugated gold nanoparticle samples in 5 mL of medium (DMEM) and sonicate to obtain uniform dispersion. The solutions thus formed are stock solutions of bare and conjugated gold nanoparticles with a concentration of 200 μg/mL.

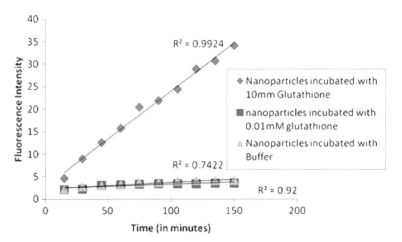

Fig. 11 Fluorescence spectra showing time-dependent release of fluorescent moiety in buffer at glutathione concentration of 10 μM and 10 mM

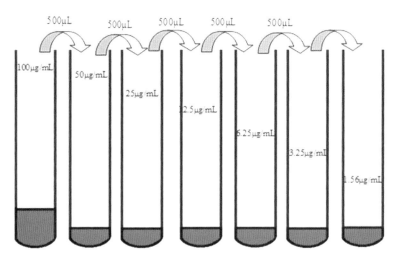

Fig. 12 Preparation of different concentrations of synthesized nanoparticles samples via serial dilution

4. Prepare various test concentrations of the bare and conjugated gold nanoparticles separately by serial dilution method from a stock solution of 200 μg/mL (*see* **Note 22** and Fig. 12).

5. The concentrations of the gold samples so obtained are 100, 50, 25, 12.5, 6.25, 3.125, and 1.56 μg/mL.

6. Add each concentration of the bare and conjugated gold nanoparticles in triplicates to the A549 cells in a 96-well plate.

7. Incubate the treated cells at 37 °C in 5 % CO_2 incubator for 24 h.

8. After incubation, add 20 μL of MTT reagent to each well and again incubate it for another 3 h at 37 °C.

9. Following incubation, add 100 μL of DMSO in each well.

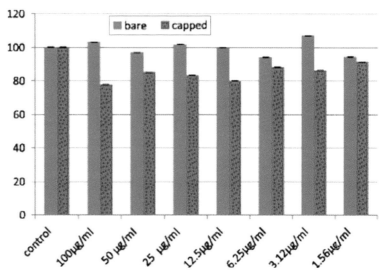

Fig. 13 Comparative analysis of the percentage cell viability of A549 cells at the various concentrations of bare and conjugated gold nanoparticles is shown. The experiment confirmed that the cells were viable at all the concentrations of the samples (ranging from 100 to 1.56 μg/mL) for bare and conjugated gold nanoparticles. At all the concentrations, the average cell viability for bare and conjugated gold nanoparticles were found to be above 95 % and 80 % respectively. The surface functionalization of the bare gold nanoparticles imparts slight incompatibility due to which the average percentage cell viability for the cells incubated with conjugated gold nanoparticles decreased in comparison to the bare gold nanoparticles. Furthermore, it was observed that while decreasing the concentration of the conjugated gold nanoparticles, a corresponding increase in the percentage cell viability occurred. This indicates that the nanovehicle is completely biocompatible at biologically relevant dosages ranging from 6.25 to 1.56 μg/mL

10. Shake the plate vigorously for about 15 min.

11. Measure the optical density (of cells) using UV spectrophotometer with a test wavelength of 570 nm and a reference wavelength of 630 nm to obtain sample signal.

12. A graphical representation of percentage cell viability verses concentration of the bare and conjugated gold nanoparticle is shown so as to assess the biocompatibility of the particles (*see* Fig. 13).

4   Notes

1. The DLS measurements of the particles are done by Photocor-FC model-1135P instrument. The light source is an argon ion laser operating at 633 nm and detection of diffracted light is measured at an angle of 90° to an incident laser beam. Diffractogram is processed by Photocor and Dynal software.

2. Vials containing the microemulsion samples are sealed with a teflon tape and are stored in cool, dry place.

3. The freeze drier instrument Ilshin model-TFD5505 is used and operated at condenser temperature −80 °C and vacuum pressure less than 5 mTorr. Before lyophilization the samples are frozen by keeping them in liquid nitrogen.

4. Before running the vacuum evaporator keep the condenser cool by circulating the chilled water. The rotation speed and the application of vacuum should be done gradually.

5. The progress and completion of the reaction is monitored through TLC method. A spot of the reaction mixture is marked on a preformed Merck TLC Plate Silica Gel 60 F254 by using a fine capillary. A mixture of EtOAc–Hexane (8:2) is used as mobile phase.

6. Vacuum desiccator with a pad of calcium chloride is used.

7. The progress and completion of the reaction is monitored through TLC method. A spot of the reaction mixture is marked on a preformed Merck TLC Plate Silica Gel 60 F254 by using a fine capillary. A mixture of CH_2Cl_2–Methanol (1:1) is used as mobile phase.

8. pH of the reaction mixture should not exceed the value of 7.2 as it may lead to lactone ring opening.

9. Use argon or nitrogen gas for maintaining inert atmosphere in the reaction system. Pass the gas by bubbling the reaction mixture slowly and continuously.

10. The progress and completion of the reaction is monitored through TLC method. A spot of the reaction mixture is marked on a preformed Merck TLC Plate Silica Gel 60 F254 by using a fine capillary. A mixture of methanol–CH_2Cl_2 (8:2) is used as mobile phase.

11. Vigorous shaking may lead to emulsion formation, so keep the separating funnel undisturbed so as to separate both phases. Remove the stopper frequently.

12. The product is stored in a desiccator with a pad of phosphorus pentoxide.

13. Dissolve 4 mg Ellman's reagent (SRL) (5,5′-Dithio-bis-(2-nitrobenzoic acid)), in 1 mL of 0.1 M sodium phosphate buffer of pH 8.0 containing 1 mM EDTA. Measure the optical absorbance of samples at 412 nm.

14. React the test sample with 1 % ninhydrin solution, 2,2-Dihydroxyindane-1,3-dione (SRL) prepared in 50 % (v/v) ethanol/water. Keep the reaction in a 95 °C water bath for 5 min and then cool down. The presence of thiol group is confirmed by appearance of purple-colored solution.

15. The product is stored in a vacuum desiccator with a pad of calcium chloride and phosphorus pentoxide in it.

16. The dialysis membrane of MWCO 12 kDa is procured form Sigma. The dialysis membrane is soaked overnight and stored at 4 °C. The dialysis membrane is rinsed thoroughly with double-distilled water for several times before use. Disperse the obtained polymeric material in minimum amount of water. Pour the material in pre-soaked dialysis bag and perform dialysis in 1,000 mL of double-distilled water for 72 h. The dialysis medium is replaced with fresh double-distilled water at every 4 h within first 24 h and at 6 h for 48–72 h.

17. The coupling reactions of the thiolated PEG with gold nanoparticles, target specific and fluorescent moieties must be done within 2–3 days of their synthesis as the thiol linkages lack stability in oxidizing environment and have strong tendency to form disulfides.

18. Gold nanoparticles are collected through liquid–liquid extraction, then collecting the aqueous layer and finally centrifuging it at $2,000–3,000 \times g$ for 5–10 min so as to obtain the nanoparticle pellet.

19. TEM analyses are performed on Technai F2 instrument operating at a voltage of 300 kV. Carbon coated copper grids are used for the sample preparation.

20. UV analysis is performed using a double-beam UV spectrophotometer instrument Systronic Model-AU 2700 (India).

21. The steady-state fluorescence measurements are carried out with a CARY Eclipse Fluorescence spectrophotometer (Varian optical spectroscopy instrument, Mulgrave, Victoria, Australia) equipped with thermostat cell holders.

22. Add 500 µL of plain medium in seven different Falcon tubes. Serially transfer 500 µL of higher concentration solution to the other tubes already containing 500 µL of plain medium with thorough mixing and sonication after every dilution.

Acknowledgment

The authors thank the Department of Science and Technology (DST), New Delhi for the financial assistance in the form of Junior Research Fellowship and the University Science Instrumentation Centre, University of Delhi, for providing the characterization facilities. The authors also thank Dr. Y. Singh (Scientist "G", IGIB, Delhi) for carrying out the cytotoxicity work in his lab.

References

1. Shenoy D, Fu W, Li J, Crasto C, Jones G, DiMarzio C, Sridhar S, Amiji M (2006) Surface functionalization of gold nanoparticles using hetero-bifunctional poly(ethylene glycol) spacer for intracellular tracking and delivery. Int J Nanomedicine 1:51

2. Gerasimov OV, Boomer JA, Qualls MM, Thompson DH (1999) Cytosolic drug delivery using pH- and light-sensitive liposomes. Adv Drug Deliv Rev 38:317–338

3. Alvarez-Lorenzo C, Bromberg L, Concheiro A (2009) Light-sensitive intelligent drug delivery systems. Photochem Photobiol 85:848–860

4. Sawahata K, Hara M, Yasunaga H, Osada Y (1990) Electrically controlled drug delivery system using polyelectrolyte gels. J Control Release 14:253–262

5. Kwon IC, Bae YH, Okano T, Kim SW (1991) Drug release from electric current sensitive polymers. J Control Release 17:149–156

6. Yuk SH, Cho SH, Lee HB (1992) Electric current-sensitive drug delivery systems using sodium alginate/polyacrylic acid composites. Pharm Res 9:955–957

7. Kost J, Leong K, Langer R (1998) Ultrasonically controlled polymeric drug delivery. Paper presented at Makromolekulare Chemie, Macromolecular Symposia 1988

8. Kost J, Leong K, Langer R (1989) Ultrasound-enhanced polymer degradation and release of incorporated substances. Proc Natl Acad Sci U S A 86:7663–7666

9. Satturwar P, Eddine MN, Ravenelle F, Leroux J-C (2007) pH-responsive polymeric micelles of poly (ethylene glycol)-b-poly (alkyl (meth) acrylate-co-methacrylic acid): influence of the copolymer composition on self-assembling properties and release of candesartan cilexetil. Eur J Pharm Biopharm 65:379–387

10. Na K, Lee KH, Bae YH (2004) pH-sensitivity and pH-dependent interior structural change of self-assembled hydrogel nanoparticles of pullulan acetate/oligo-sulfonamide conjugate. J Control Release 97:513–525

11. Chen S-C, Wu Y-C, Mi F-L, Lin Y-H, Yu L-C, Sung H-W (2004) A novel pH-sensitive hydrogel composed of N,O-carboxymethyl chitosan and alginate cross-linked by genipin for protein drug delivery. J Control Release 96:285–300

12. Hrubý M, Koňák Č, Ulbrich K (2005) Polymeric micellar pH-sensitive drug delivery system for doxorubicin. J Control Release 103:137–148

13. Ito Y, Casolaro M, Kono K, Imanishi Y (1989) An insulin-releasing system that is responsive to glucose. J Control Release 10:195–203

14. Shiino D, Murata Y, Kataoka K, Koyama Y, Yokoyama M, Okano T, Sakurai Y (1994) Preparation and characterization of a glucose-responsive insulin-releasing polymer device. Biomaterials 15:121–128

15. Hisamitsu I, Kataoka K, Okano T, Sakurai Y (1997) Glucose-responsive gel from phenylborate polymer and poly(vinyl alcohol): prompt response at physiological pH through the interaction of borate with amino group in the gel. Pharm Res 14:289–293

16. Dong-June C, Yoshihiro I, Yukio I (1992) An insulin-releasing membrane system on the basis of oxidation reaction of glucose. J Control Release 18:45–53

17. Ulijn RV (2006) Enzyme-responsive materials: a new class of smart biomaterials. J Mater Chem 16:2217–2225

18. Thornton PD, McConnell G, Ulijn RV (2005) Enzyme responsive polymer hydrogel beads. Chem Commun 5913–5915

19. Toledano S, Williams RJ, Jayawarna V, Ulijn RV (2006) Enzyme-triggered self-assembly of peptide hydrogels via reversed hydrolysis. J Am Chem Soc 128:1070–1071

20. Miyata T, Asami N, Uragami T (1999) Preparation of an antigen-sensitive hydrogel using antigen-antibody bindings. Macromolecules 32:2082–2084

21. Lu ZR, Kopečková P, Kopeček J (2003) Antigen responsive hydrogels based on polymerizable antibody Fab′ fragment. Macromol Biosci 3:296–300

22. Zhang R, Bowyer A, Eisenthal R, Hubble J (2007) A smart membrane based on an antigen-responsive hydrogel. Biotechnol Bioeng 97:976–984

23. Koo AN, Lee HJ, Kim SE, Chang JH, Park C, Kim C, Park JH, Lee SC (2008) Disulfide-cross-linked PEG-poly (amino acid)s copolymer micelles for glutathione-mediated intracellular drug delivery. Chem Commun 6570–6572

24. Tsarevsky NV, Matyjaszewski K (2005) Combining atom transfer radical polymerization and disulfide/thiol redox chemistry: a route to well-defined (bio)degradable polymeric materials. Macromolecules 38:3087–3092

25. Aerry S, De A, Kumar A, Saxena A, Majumdar D, Mozumdar S (2012) Synthesis and characterization of thermoresponsive copolymers for drug delivery. J Biomed Mater Res A 101A:2015–2026

26. Bae YH, Okano T, Hsu R, Kim SW (1987) Thermo-sensitive polymers as on-off switches for drug release. Makromol Chem Rapid Commun 8:481–485

27. Chung JE, Yokoyama M, Yamato M, Aoyagi T, Sakurai Y, Okano T (1999) Thermo-responsive drug delivery from polymeric micelles constructed using block copolymers of poly(N-isopropylacrylamide) and poly(butylmethacrylate). J Control Release 62:115–127

28. Li Y, Pan S, Zhang W, Du Z (2009) Novel thermo-sensitive core-shell nanoparticles for targeted paclitaxel delivery. Nanotechnology 20:065104

29. Hong R, Han G, Fernández JM, Kim B, Forbes NS, Rotello VM (2006) Glutathione-mediated delivery and release using monolayer protected nanoparticle carriers. J Am Chem Soc 128:1078–1079

30. Jones DP, Carlson JL, Mody VC Jr, Cai J, Lynn MJ, Sternberg P Jr (2000) Redox state of glutathione in human plasma. Free Radic Biol Med 28:625–635

31. Anderson ME (1998) Glutathione: an overview of biosynthesis and modulation. Chem Biol Interact 111:1–14

32. Chompoosor A, Han G, Rotello VM (2008) Charge dependence of ligand release and monolayer stability of gold nanoparticles by biogenic thiols. Bioconjug Chem 19:1342–1345

33. Han G, Chari NS, Verma A, Hong R, Martin CT, Rotello VM (2005) Controlled recovery of the transcription of nanoparticle-bound DNA by intracellular concentrations of glutathione. Bioconjug Chem 16:1356–1359

34. Li D, Li G, Guo W, Li P, Wang E, Wang J (2008) Glutathione-mediated release of functional plasmid DNA from positively charged quantum dots. Biomaterials 29:2776–2782

35. Liu J, Pang Y, Huang W, Huang X, Meng L, Zhu X, Zhou Y, Yan D (2011) Bioreducible micelles self-assembled from amphiphilic hyperbranched multiarm copolymer for glutathione-mediated intracellular drug delivery. Biomacromolecules 12:1567–1577

36. Sun H, Guo B, Cheng R, Meng F, Liu H, Zhong Z (2009) Biodegradable micelles with sheddable poly(ethylene glycol) shells for triggered intracellular release of doxorubicin. Biomaterials 30:6358–6366

37. Tang L-Y, Wang Y-C, Li Y, Du J-Z, Wang J (2009) Shell-detachable micelles based on disulfide-linked block copolymer as potential carrier for intracellular drug delivery. Bioconjug Chem 20:1095–1099

38. Yang P-H, Sun X, Chiu J-F, Sun H, He Q-Y (2005) Transferrin-mediated gold nanoparticle cellular uptake. Bioconjug Chem 16:494–496

39. Han G, Ghosh P, Rotello VM (2007) Functionalized gold nanoparticles for drug delivery. Nanomedicine 2:113–123

40. Pingarrón JM, Yáñez-Sedeño P, González-Cortés A (2008) Gold nanoparticle-based electrochemical biosensors. Electrochim Acta 53:5848–5866

41. El-Sayed IH, Huang X, El-Sayed MA (2005) Surface plasmon resonance scattering and absorption of anti-EGFR antibody conjugated gold nanoparticles in cancer diagnostics: applications in oral cancer. Nano Lett 5:829–834

42. Shukla S, Priscilla A, Banerjee M, Bhonde RR, Ghatak J, Satyam P, Sastry M (2005) Porous gold nanospheres by controlled transmetalation reaction: a novel material for application in cell imaging. Chem Mater 17:5000–5005

43. Chen J, Saeki F, Wiley BJ, Cang H, Cobb MJ, Li Z-Y, Au L, Zhang H, Kimmey MB, Li X (2005) Gold nanocages: bioconjugation and their potential use as optical imaging contrast agents. Nano Lett 5:473–477

44. Cheng MM-C, Cuda G, Bunimovich YL, Gaspari M, Heath JR, Hill HD, Mirkin CA, Nijdam AJ, Terracciano R, Thundat T (2006) Nanotechnologies for biomolecular detection and medical diagnostics. Curr Opin Chem Biol 10:11–19

45. Rosi NL, Mirkin CA (2005) Nanostructures in biodiagnostics. Chem Rev 105:1547–1562

46. Pissuwan D, Niidome T, Cortie MB (2011) The forthcoming applications of gold nanoparticles in drug and gene delivery systems. J Control Release 149:65–71

47. Ghosh P, Han G, De M, Kim CK, Rotello VM (2008) Gold nanoparticles in delivery applications. Adv Drug Deliv Rev 60:1307–1315

48. Gu Y-J, Cheng J, Lin C-C, Lam YW, Cheng SH, Wong W-T (2009) Nuclear penetration of surface functionalized gold nanoparticles. Toxicol Appl Pharmacol 237:196–204

49. Thomas M, Klibanov AM (2003) Conjugation to gold nanoparticles enhances polyethylenimine's transfer of plasmid DNA into mammalian cells. Proc Natl Acad Sci U S A 100:9138–9143

50. Zhang G, Yang Z, Lu W, Zhang R, Huang Q, Tian M, Li L, Liang D, Li C (2009) Influence of anchoring ligands and particle size on the colloidal stability and in vivo biodistribution of polyethylene glycol-coated gold nanoparticles in tumor-xenografted mice. Biomaterials 30:1928–1936

51. Bhattacharya R, Patra CR, Earl A, Wang S, Katarya A, Lu L, Kizhakkedathu JN, Yaszemski MJ, Greipp PR, Mukhopadhyay D (2007) Attaching folic acid on gold nanoparticles using noncovalent interaction via different polyethylene glycol backbones and targeting of

cancer cells. Nanomed Nanotechnol Biol Med 3:224–238

52. Garg S, De A, Nandi T, Mozumdar S (2013) Synthesis of a smart gold nano-vehicle for liver specific drug delivery. AAPS PharmSciTech 14:1219–1226

53. Garg S (2013) Development of nano-particulate systems for their applications in bio-medical area. Ph.D. Thesis submitted at the Department of Chemistry, University of Delhi

54. Kobayashi K, Sumitomo H, Ina Y (1985) Synthesis and functions of polystyrene derivatives having pendant oligosaccharides. Polym J 17:567–575

55. Besson T, Coudert G, Guillaumet G (1991) Synthesis and fluorescent properties of some heterobifunctional and rigidized 7-aminocoumarins. J Heterocycl Chem 28:1517–1523

56. Furniss PS, Hannaford AJ, Smith PWG, Tatchell AR (2006) Vogel's textbook of practical organic chemistry, 5th edn. Pearson Education, New Delhi

57. Stefanko MJ, Gun'ko YK, Rai DK, Evans P (2008) Synthesis of functionalised polyethylene glycol derivatives of naproxen for biomedical applications. Tetrahedron 64:10132–10139

58. Ladd DL, Henrichs PM (1998) Synthesis and NMR characterization of monomethoxypoly(ethylene glycol) aldehydes from monomethoxypoly(ethylene glycol) tosylates. Synth Commun 28:4143–4149

Chapter 15

Intranasal Delivery of Chitosan–siRNA Nanoparticle Formulation to the Brain

Meenakshi Malhotra, Catherine Tomaro-Duchesneau, Shyamali Saha, and Satya Prakash

Abstract

Neurodegeneration is characterized by a progressive loss of neuron structure and function. Most neurodegenerative diseases progress slowly over the time. There is currently no cure available for any neurodegenerative disease, and the existing therapeutic interventions only alleviate the symptoms of the disease. The advances in the drug discovery research have come to a halt with a lack of effective means to deliver drugs at the targeted site. In addition, the route of delivering the drugs is equally important as most invasive techniques lead to postoperative complications. This chapter focuses on a non-invasive, intranasal mode of therapeutic delivery using nanoparticles, which is currently being explored. The intranasal route of delivery is a well-established route to deliver drugs via the olfactory and trigeminal neuronal pathways. It is known to be the fastest and most effective way to bypass the blood–brain barrier to reach the central nervous system. The presented chapter highlights the method of intranasal delivery in mice using chitosan–siRNA nanoparticle formulation, under mild anesthesia and the identification of successful siRNA delivery in the brain tissues, through histology and other well-established laboratory protocols.

Key words Intranasal, Brain, Olfactory, Nanoparticles, siRNA

1 Introduction

Therapeutic delivery to the brain is of great importance and presents a major challenge, for the treatment of neurodegenerative diseases (NDDs) such as Alzheimer's Parkinson's, multiple sclerosis, Huntington's, Amyotrophic lateral sclerosis, spinocerebellar ataxia, etc. NDDs substantially affect patients' quality of life, and many limit the life span of an individual following disease onset [1]. Though the current research has focused on understanding the mechanism(s) and pathological pathways underlying the various NDDs and their associated drug targets, delivery of the therapeutic molecules at the targeted site remains a major issue [2]. Following the identification of potential therapeutic targets, there is a dire need to enhance the therapeutic molecule's bioavailability and

Kewal K. Jain (ed.), *Drug Delivery System*, Methods in Molecular Biology, vol. 1141,
DOI 10.1007/978-1-4939-0363-4_15, © Springer Science+Business Media New York 2014

pharmacokinetic profiles [3]. In terms of therapeutic delivery to the brain, the main obstacle is the blood–brain barrier (BBB), formed by a network of brain endothelial cells connected by tight junctions [4]. This membranous structure creates a selective barrier limiting the entry of neuroprotective therapeutics. To overcome this barrier, various invasive techniques including convection-enhanced diffusion, intrathecal/intraventricular and intracerebral drug delivery systems are used, which lead to a mechanical breach of the BBB [5–8]. In contrast to the existing invasive techniques, the intranasal route of delivery has emerged as a non-invasive, direct nose-to-brain delivery route, which has been widely used in rodent animals. Unfortunately, the exact mechanism of therapeutic uptake and the translatability of the route in humans are yet to be elucidated [9].

The intranasal route of delivery allows the therapeutic to bypass the BBB, avoiding the elimination of therapeutic molecules by liver, kidney filtration, gastrointestinal tract, and serum degradation [10]. The intranasal route leading to the central nervous system (CNS) can be classified in two pathways. (1) The intracellular axonal transport, which involves internalization of the therapeutics by the olfactory receptor neurons present in the olfactory mucosa, followed by *trans*-synaptic transport to secondary neurons via receptor-mediated and non-receptor-mediated endocytosis, following either an anterograde or a retrograde transport, leading its way through the olfactory bulb and distribution to other areas of the CNS [11]. In addition to olfactory transport, the trigeminal nerve also plays a role in transporting therapeutics via the intraneuronal route [11]. (2) The second pathway is the extracellular transport, a comparatively faster route to transport therapeutics to the brain. In this pathway, the therapeutics are absorbed across the olfactory epithelium through the Bowman's gland cells either via receptor-mediated endocytosis or passive diffusion or by following a paracellular pathway between the tight junctions and entering the perineural channels surrounding the olfactory and trigeminal nerves [12]. Figure 1 illustrates the extraneuronal pathway, followed by the therapeutics to reach CNS on intranasal delivery via olfactory and trigeminal nerves.

Most NDDs are known to result from gene–environment interactions. The mutational events such as deletion, insertion, polymorphisms, and inversions may produce a dominant defective feature, which is inheritable and can lead to genetically inherited diseases [13]. The internal safety mechanisms of a cell check for such unintentional lethalities. One such mechanism is RNA interference (RNAi), which has gained a lot of importance in current research due to its potential in therapeutic applications. RNAi is a mechanism, which is mediated by short interfering RNAs (siRNAs) of 18–21 bp in length, that offer specific gene silencing at the mRNA level, followed by protein inhibition [14].

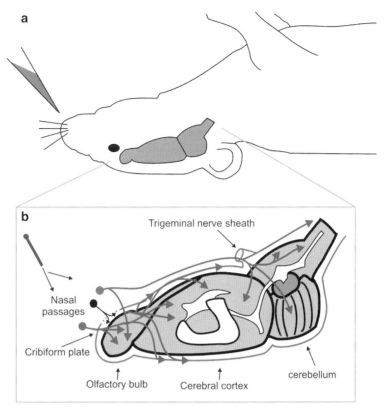

Fig. 1 (**a**) The head-back supine position of the mouse under mild gas (Isoflurane) anesthesia, required for intranasal delivery of the nanoparticle formulation. (**b**) Anatomical representation of the extraneuronal pathway, which facilitates the uptake and delivery of the therapeutic formulation into the brain via olfactory nerves (shown in *red*) and trigeminal nerves (shown in *blue*). This figure is adapted and modified from Thorne et al. [11] (Color figure online)

Successful delivery of siRNA, in vivo, has been a challenge due to the molecule's transient nature, with high susceptibility to enzymatic degradation, low cellular uptake, rapid clearance from the blood, and off-target effects [15]. The clinical application of siR-NAs for CNS delivery has been the most challenging task, though various techniques that involve chemical modifications of siRNAs and complexation with transfection agents have been employed to increase the stability and cellular uptake issues of siRNAs [16]. Thus, there is a need for a delivery vehicle that successfully complexes the siRNA into nanoparticles and can make its way via extraneuronal/intraneuronal pathways of intranasal route, combating the enzymatically active, low pH of nasal epithelium, and deliver siRNA at the targeted site, without compromising its efficiency. In recent years, various nanoparticles such as polyplexes, dendriplexes, and exosomes have shown success with in vivo siRNA delivery [17–19]. Table 1 represents a comprehensive list

Table 1
Nanoparticle formulations used for intranasal delivery of different payloads to the brain, using in vivo studies

Nanoparticle type	Payload	Mean particle size (nm)	Brain region	Function	References
Wheat germ agglutinin–PEG–PLA	Coumarin	<100	OB, OT, CR, CL	Brain delivery	[38]
Chitosan	Estradiol	200–300	CSF	Alzheimer's disease therapeutic	[39]
Wheat germ agglutinin–PEG–PLA	Vasoactive intestinal peptide	100–120	OB, OT, CR, CL	Brain delivery	[40]
Ulex europaeus agglutinin I–PEG–PLA	6-Coumarin	100–150	OB, OT, CR, CL	Brain delivery	[41]
Rutile/Anatase TiO_2	N/A	80 and 155	CR, TH, HPP	Brain delivery	[42, 43]
Protamine–PEG–PLA	Coumarin	80–90	OB, OT, CR, CL	Brain delivery	[44]
Wheat germ agglutinin–PEG–PLA	Quantum dots	50–150	OB, OT, CR, HPP	Brain imaging	[45]
Odoranlectin–PEG–PLA	Urocortin peptide	<120	Olfactory epithelium, cerebral cortex	Parkinson's disease therapeutic	[46]
Methoxy PEG–PLA	Nimodipine	<100	CSF, OB, OT, CR, CL	Brain delivery	[47]
Polylactic acid	Neurotoxin-I	65	OB	Brain delivery	[48]
Chitosan	Thymoquinone	150–200	NS	Brain delivery	[49]
Chitosan	Bromocriptine	<200	NS	Parkinson's disease	[50]
Chitosan and Pluronic PF127 gel	Levodopa	164.5	OB, OT, CR, CL	Parkinson's disease therapeutic	[51]

Polymeric N-isopropyl acryl amide	Curcuminoids, demethoxycurcumin, bisdemethoxycurcumin	<100	CR	Cerebral ischemia/MCAO therapeutic	[52]
Chitosan	Carnosic acid	180–350	HPP	Brain delivery	[53]
Chitosan	Rivastigmine	<200	CSF, OB	Alzheimer's disease therapeutic	[54]
Solid lipid	Ondansetron HCl	320–498	NS	Brain delivery	[55]
Iron oxide	N/A	400–500	OB, Striatum HPP, CR, CL	Neurotoxicity	[56]
Chitosan–PEG–TAT/MGF peptide	Scrambled siRNA	<20	CR, CL	Brain delivery	[35]
Lactoferrin PEG–PCL	6-Coumarin	<100	OB, OT, CR, CL, HPP	Alzheimer's disease therapeutic	[57]
Cationic liposomes	Ovalbumin	270–320	Cerebral cortex	Brain delivery	[58]
Exosomes	Exo-cur/ExoJSI124	30–100	Microglial cells	Brain inflammatory disease therapeutic	[59]
Lipid	Valproic acid	140–180	OB	Brain delivery	[60]
Poly(amidoamine) G7 dendrimers	siRNAs	150	OB, HPT, CR, CL	Brain delivery	[18]

Abbreviations: OB olfactory bulb, *OT* olfactory tract, *CR* cerebrum, *CL* cerebellum, *HPP* hippocampus, *HPT* hypothalamus, *CSF* cerebrospinal fluid, *TH* thalamus, *NS* nonspecific, *siRNA* small interfering RNA's, *PLA* poly lactic acid, *TAT* transactivated transcription factor, *MGF* mechano growth factor, *PCL* poly(ε-caprolactone)

of various nanoparticle formulations that have specifically been used for intranasal delivery of various payloads (drugs/gene/peptides) to the CNS.

Cationic polymeric nanoparticles provide an advantage for the delivery of therapeutics, as they are easily taken up by the cells via adsorptive-mediated endocytosis. In this chapter, we focus on the chitosan polymer, which has been widely used to deliver nerve growth factors, insulin and drugs to the brain via an intranasal route of delivery [20–22]. Chitosan polymer has various advantages over other polymeric nanoparticles, (1) it is biodegradable, derived from chitin, which is obtained from the shells of shrimps and crabs [23, 24]. (2) The presence of hydroxyl group at the C6 position and the amine group at the C2 position of its repeating units allows for chemoselective conjugations to other moieties [25, 26]. (3) The presence of amine groups helps complex with negatively charged oligonucleotides to form nanoparticles. (4) The polymer exhibits a "proton sponge" effect that refers to the swelling behavior of the polymer on encountering an acidic pH inside the cell's endosome, thereby making the polymer an efficient carrier for therapeutic molecules [27]. (5) Chitosan is known to be a mucoadhesive agent; the amines in chitosan react with sialic residues present on the mucosal layer, reducing their clearance rate from the nasal cavity [28]. Due to its mucoadhesive property, it has been used for intranasal delivery of various formulations for ocular and pulmonary diseases [29–33]. In this chapter, we have utilized the ability of surface-functionalized chitosan nanoparticles to deliver siRNA to the brain, following an intranasal route.

2   Materials

2.1   Stock Preparation

1. Chitosan (cationic polymer) stock solution: 0.05 % (w/v) of low molecular weight chitosan (10 kDa) is dissolved in 1 % (v/v) acetic acid solution prepared in ddH$_2$O. Filter the solution with 0.2 μm filter prior to use.

2. Sodium tripolyphosphate (TPP) (anionic cross-linker) stock solution: 0.07 % (w/v) of TPP is dissolved in ddH$_2$O and pH is adjusted to 3 using HCl. Filter the solution with 0.2 μm filter prior to use.

3. siRNA stock solution: scrambled (non-targeting) biotin-tagged siRNA obtained from Dharmacon in a lyophilized vial is dissolved in RNase and DNase-free water to get a required stock concentration of siRNA needed for the experiment.

2.2   Preparation of Chitosan–siRNA Nanoparticles

1. 10 ml glass vials with screw caps, autoclaved at 121 °C for 15 min.

2. 20 μl, 200 μl, and 1 ml micropipettes and autoclaved micropipette tips.

3. Magnetic stirrer and magnetic stir bars of 13×8 mm dimension.

4. Centrifuge.

5. Amicon Ultra Centrifugal Filter Units (Millipore).

6. Autoclaved eppendorf tubes, micropipettes, and pipette tips.

2.3   Intranasal Delivery of Chitosan– siRNA Formulation to Mice

1. Cocktail anesthetic for mice: Mix 1 ml of Ketamine (100 mg/ ml), 0.5 ml of Xylazine (20 mg/ml), and 0.3 ml of Acepromazine (10 mg/ml). Add 8.2 ml of sterile saline (0.85 % (w/v) NaCl) to adjust the total volume to 10 ml. Store cocktail in dark place, replace every 6 months (*see* **Note 1**).

2. Gas anesthesia machine with filter.

3. Transparent Induction chamber (glass/polycarbonate).

4. Isoflurane.

5. Microwave, microwavable heating pads (Snuggle safe).

6. Micropipettes and autoclaved pipette tips.

2.4   In Vivo Histology and Tissue Toxicity Analysis

1. Xylene (concentrated, >98.5 %), Ethanol (70, 90, 100 % (v/v)), prepared in deionized (DI) water, 3 % (v/v) Hydrogen peroxide (H_2O_2).

2. 1× Phosphate Buffer Saline (PBS) solution: 8 g of NaCl, 0.2 g of KCl, 1.44 g of Na_2HPO_4, and 0.2 g of KH_2PO_4 to 800 ml of double-distilled/deionized water, and adjust pH to 7.4 using HCl. Sterilize by autoclaving at 121 °C for 15 min.

3. Fixative: 4 % (w/v) paraformaldehyde prepared in PBS.

4. VECTASTAIN® ABC Kit: in a mixing bottle provided by the manufacturer, mix 5 ml of PBS, with two drops of Reagent A and two drops of Reagent B and mix immediately. Let the solution stand for 30 min prior to use (*see* **Note 2**).

5. Vector® DAB Substrate (3,3′-diaminobenzidine): In a 15 ml centrifuge tube, add 5 ml of DI water, two drops of buffer stock, four drops of DAB stock and two drops of H_2O_2 (*see* **Note 3**).

6. Hematoxylin (Fisher Scientific) and vector mount (permanent mounting medium).

7. DeadEnd™ Colorimetric TUNEL assay kit (Promega). Stock preparation:

   (a) Prepare 20 µg/ml from 10 mg/ml of Proteinase K stock solution, by diluting it 1:500 in PBS.

   (b) Prepare rTDT reaction mix for a standard calculation of 100 µl/reaction: [(98 µl of Equilibration Buffer + 1 µl of Biotinylated nucleotide mix + 1 µl of rTDT mix) × number of reactions].

   (c) Prepare 2× from 20× SSC stock by diluting to 1:10.

8. Coplin glass jars with lids for staining.

9. Humidified chambers for microscope slides.

10. 37 °C incubator.

11. Glass coverslips.

12. Micropipettes and pipette tips.

3   Methods

3.1   Preparation of Chitosan–siRNA Nanoparticle Formulation

The preparation of the nanoparticle formulation for siRNA delivery is dependent on the targeted site and route of administration. These factors are important in deciding what size and surface charge is ideal for the nanoparticle formulation. In addition, to make the nanoparticles specifically targeted, they can be surface graphed with peptides, ligands or antibodies. We have previously prepared nanoparticle formulations with various methods for three different applications, with different targeting peptides, by organically synthesizing the derivatized forms of chitosan polymer [25, 26, 34–36]. However, the basic method of forming polyplex nanoparticles with siRNA via the ionic gelation method remains the same. The following method details the procedure common to the preparation of polyplex nanoparticles-siRNA with or without the derivatized chitosan polymer. We encourage readers to review the preparation of derivatized forms of chitosan in our previously published articles.

1. The scrambled biotin-tagged siRNA (2 μg) is premixed with 200 μl of TPP, 0.07 % (w/v) (pH 3) and dropped into the 800 μl of chitosan polymer solution, 0.05 % (w/v), under constant magnetic stirring at 800 rpm for 1 h [37].

2. The weight ratio of chitosan–siRNA optimized in our previous study was 200:1 [37], determined by performing gel retardation assay (*see* **Note 4**).

3. Nanoparticles, prepared at an optimal weight ratio, should be characterized for size and surface charge through dynamic light scattering (DLS) and zeta potential techniques, respectively. The nanoparticles optimized in our previous studies were <20 nm in size with a surface charge of ±16 mV [37].

4. The optimal conditions achieved for nanoparticle–siRNA formulation should be tested in vitro on a desired cell line to achieve maximum transfection and minimum cytotoxicity (*see* **Note 5**).

5. For the in vivo study, the siRNA dose of 0.5 mg/kg of animal weight was optimized by our group [35]. Thus, the nanoparticles are prepared in bulk volumes at initial concentrations as described in **step 1** and later concentrated down to a volume of 30 μl (for intranasal administration), using Amicon Ultra Centrifugal Filter Units (Amicon), with a molecular weight cut-off of 3 kDa.

3.2 Intranasal Delivery of Chitosan–siRNA Formulation to Mice

The intranasal delivery of a nanoparticle formulation in animals can be performed using different methods. If the formulation needs a single administration, the standard anesthesia cocktail can first be administered to the animals via intraperitoneal injection. The cocktail anesthesia is a strong sedative and ensures that the animals are unconscious for 15–20 min. However, this may not be the ideal form of anesthesia for a long term study as the strong anesthetic itself can lead to toxicity and mortality. In long-term studies, isoflurane gas anesthesia is preferred as a milder sedative with animal recovery within a couple of minutes. The following procedure details the isoflurane mediated gas anesthesia method for intranasal delivery of nanoparticle formulation.

1. Place the animal in the induction chamber and adjust the oxygen flow meter to 0.8–1.5 l/min.

2. Adjust the isoflurane vaporizer to 3–5 % to completely sedate the animal.

3. Once the animal is completely sedated, quickly open the chamber and place the animal in a head back position and close. Monitor the breathing of the animal (*see* **Note 6**).

4. Once the breathing of the animal is at a steady rate, adjust and maintain the oxygen flow meter to 400–800 ml/min and the isoflurane vaporizer at 2–2.5 %.

5. During this maintenance period, administer a total of 30 μl of the nanoparticle-siRNA formulation intranasally to the animal as 5 μl/drop over a period of 15–20 min (*see* **Note 7**).

6. Turn off the isoflurane vaporizer but keep the animal on oxygen for a few minutes. As the animal starts to move transfer it to the cage and allow it to fully recover.

3.3 In Vivo Histology Analysis

The VECTASTAIN® ABC Kit employs avidin/biotin technology, which is highly active and uniform. The two reagents in the kit: Reagent A has Avidin DH and Reagent B has the biotinylated horseradish peroxidase (HRP) enzyme. The avidin has affinity for biotin, which allows the formation of macromolecular complexes, which are stable for several hours after formation. The advantage of using VECTASTAIN® ABC Reagent is to detect the scrambled siRNA, which has a biotin tag. Vector® DAB Substrate is used as a chromogen that produces a brown reaction in the presence of HRP (*see* **Note 8**).

1. Deparaffinize the tissue slides by sequentially immersing them in three separate fresh xylene jars for 3 min each.

2. Rinse the tissue slides two times, sequentially in 100 % (v/v) ethanol jar for 3 min each.

3. Hydrate the tissue slides sequentially for 3 min in 90 % (v/v) ethanol (two times) and 70 % (v/v) ethanol (one time), followed by rinse in tap water for 5 min.

4. In a humidified chamber, incubate the tissue slides with 3 % (v/v) H_2O_2 prepared in water for 10 min (*see* **Note 9**).

5. Wash the tissue slides in PBS buffer for 5 min.

6. In a humidified chamber, incubate the tissue slides for 30 min with VECTASTAIN® ABC Reagent (*see* **Note 2**).

7. Wash the tissue slides in PBS buffer for 5 min.

8. Incubate the tissue slides in peroxidase substrate solution (DAB) for 10 min (*see* **Note 3**).

9. Rinse the tissue slides in tap water.

10. Counterstain the tissue slides with Hematoxylin, for 30 s, followed by quick rinsing dips in DI water.

11. Dehydrate the tissue slides serially for 2 min each in 70, 95, and 100 % (v/v) ethanol jars, followed by two jars of xylene.

12. Mount the tissue sections using Permount solution and let dry, before observing under a microscope (*see* **Note 10**).

3.4 Toxicity Analysis

The DeadEnd™ Colorimetric TUNEL (*T*dT-mediated d*U*TP *N*ick-*E*nd *L*abeling) assay is a method to detect DNA fragmentation by incorporating the biotinylated nucleotide at the 3′-OH of the nicked DNA of apoptotic cells. The incorporation of nucleotides is catalyzed by the enzyme Terminal Deoxynucleotidyl Transferase, Recombinant (rTdT). Streptavidin HRP then binds to the biotinylated nucleotides, which are detected using the peroxidase substrate, DAB. This procedure stains the apoptotic nuclei dark brown in color.

1. Deparaffinize the tissue slides by sequentially immersing them in three separate fresh xylene jars for 3 min each.

2. Rinse the tissue slides two times, sequentially in 100 % (v/v) ethanol jar for 5 min and 3 min respectively.

3. Hydrate the tissue slides serially in 95, 85, 70, and 50 % (v/v) ethanol jars for 3 min each.

4. Rinse the tissue slides in 0.85 % (w/v) NaCl solution for 3 min, followed by PBS wash for 5 min.

5. In a humidified chamber, incubate the tissue slides with Streptavidin–HRP enzyme (1:500 prepared in PBS) for 15 min, followed by a rinse in PBS for 5 min and again incubate the tissue slides with 3 % (v/v) H_2O_2 prepared in water for 10 min, followed by a rinse in PBS for 5 min (*see* **Note 11**).

6. Fix the tissue slides in 4 % (w/v) PFA for 15 min, followed by rinse in PBS (two times) for 5 min each.

7. Remove excess liquid from the tissue slides and incubate them in a humidified chamber with 100 μl of 20 μg/ml of Proteinase K solution (*see* **Note 12**).

8. Rinse the tissue slides in PBS (two times) for 5 min each.

9. Refix the tissue slides by immersing them in 4 % (w/v) PFA for 15 min, followed by rinse in PBS (two times) for 5 min each.

10. Remove excess liquid from the tissue slides and incubate them in a humidified chamber with 100 μl of Equilibration Buffer (provided in the kit) for 10 min.

11. Incubate the tissue slides with 100 μl of rTDT reaction mix for 60 min in a humidified incubator maintained at 37 °C, as prepared in **item 7b** of Subheading 2.4. Cover the slides with plastic coverslips, provided in the kit (*see* **Note 13**).

12. Incubate the tissue slides in a humidified chamber at room temperature with 2× SSC solution for 15 min, as prepared in **item 7c** of Subheading 2.4 (*see* **Note 14**).

13. Rinse the tissue slides in PBS solution (three times) for 5 min each.

14. In a humidified chamber, incubate the tissue slides with Streptavidin–HRP enzyme (1:500 prepared in PBS) for 30 min.

15. Rinse the tissue slides in PBS solution (three times) for 5 min each.

16. Incubate the tissue slides with freshly prepared DAB solution (as explained in Subheading 2.4, **item 5**) 100 μl per reaction for 10 min at room temperature.

17. Rinse the tissue slides with DI water to remove excess stain from the tissue slides.

18. Counterstain the tissue slides with Hematoxylin, for 30 s, followed by quick rinsing dips in DI water.

19. Dehydrate the tissue slides serially for 2 min each in 70, 95, and 100 % (v/v) ethanol jars, followed by two jars of xylene.

20. Mount the tissue sections using Permount solution and let dry, before observing under the microscope (*see* **Note 10**).

4   Notes

1. The mixed cocktail should be stored in a dark, cool place and should not be stored for more than 6 months. Administer 0.1 ml/10 g animal body weight intraperitoneally. Always remember to tap the bubbles out while loading the anesthetic, or saline into the syringe.

2. The ABC VECTASTAIN Elite solution should be prepared fresh every time and not stored for long periods of time. Make sure to let the solution stand for 30 min after preparation, prior to use.

3. The DAB solution should be prepared just prior to use, it should be protected from light and used within 30 min of its preparation.

4. The gel retardation assay can be performed on a 4 % (w/v) agarose gel, placed in an electrophoretic unit. The samples with various chitosan:siRNA weight ratios are loaded onto the gel, premixed with 1:6 dilution of the 6× orange dye. The gel should run for 4 h at 55 V in Tris–borate EDTA (TBE) buffer (pH 8.3). The TBE buffer should be premixed with ethidium bromide at a concentration of 0.5 μg/ml required for the visualization of RNA bands under UV transillumination at 365 nm.

5. A fluorescently tagged scrambled siRNA (e.g., Cy-5 labeled scrambled siRNA) can be complexed with chitosan polymer to determine the optimal transfection of the nanoparticle-siRNA complex, in vitro and quantified on cells using a microtiter plate spectrophotometer at 660 nm. Cytotoxicity can be analyzed using a standard CellTiter 96® AQueous One Solution Cell Proliferation Assay (MTS) assay, as per the manufacturer's protocol (Promega).

6. Prior to intranasal administration of the therapeutic dose, it is important to monitor the breathing of the animal and ensure that it is at a steady and relaxed rate.

7. Before administering the drop intranasally, make sure that the animal is in head back position with its head, parallel to the ground surface. After administering each dose of 5 μl, give a break of 3–4 min to ensure that the animal takes the drop in and is breathing at a steady rate. Do not block both the nostrils with a dose at the same time, as it can lead to loss of breath and mortality.

8. For histology analysis, it is preferred to perfusion fix the animal using 4 % (w/v) paraformaldehyde (PFA). This can only be used if the organs are not being used for any other analysis, such as protein or RNA analysis. The harvested organs should be kept at 4 °C in 4 % (w/v) PFA for 48 h. The tissues should be trimmed to 3 mm thick sections and stored in 70 % (v/v) ethanol in histology cassettes for at least 24 h before paraffin embedding. The paraffin-embedded tissues should be processed into no less than 4 μm thick sections on slides for histology analysis.

9. Incubation with 3 % (v/v) H_2O_2 is important at this step to quench the endogenous peroxidase activity in red blood cells (RBCs), which gives a false positive result.

10. Mounting of the tissue sections is tricky and should be done with care to avoid the formation of bubbles between the tissue sections and coverslip. Place few drops of Permount solution on and beside the tissue section on the slide. Slowly bend the coverslip on the edge of the slide making a 30–45° angle and as soon as the coverslip touches the solution, slowly and lightly slide the coverslip over the tissue section.

11. Incubation with 3 % (v/v) H_2O_2 is important to perform prior to TUNEL assay, in order to quench the biotin-tag on scrambled siRNA, which may give a false positive result.

12. Prolonged incubation with Proteinase K can cause the tissue sections to come off the slides as it permeabilizes the tissue sections. Thus, paraffin sections of 5–10 μm should be incubated for a shorter duration. The user can optimize the duration based on their tissue slide sections.

13. The rTDT reaction mix should be prepared fresh as mentioned in Subheading 2.4, **item 7b**. This step allows the end-labeling reaction to occur on the fragment DNA of apoptotic cells.

14. The SSC reaction is performed to terminate the end-labeling reaction. It is important to ensure that there are no precipitated salts present before diluting the SSC solution from 20× to 2×. If so, thaw and mix the solution well before making a working stock solution of 2×.

Acknowledgement

We gratefully acknowledge the assistance Canadian Institute of Health Research (CIHR) to Dr. S. Prakash and the support of FRSQ doctoral scholarship to M. Malhotra and an NSERC Alexander Graham Bell Graduate doctoral scholarship to C. Tomaro-Duchesneau.

References

1. Barchet TM, Amiji MM (2009) Challenges and opportunities in CNS delivery of therapeutics for neurodegenerative diseases. Expert Opin Drug Deliv 6:211–225

2. Brasnjevic I, Steinbusch HWM, Schmitz C, Martinez-Martinez P (2009) Delivery of peptide and protein drugs over the blood–brain barrier. Prog Neurobiol 87:212–251

3. Lansbury PT Jr (2004) Back to the future: the 'old-fashioned' way to new medications for neurodegeneration. Nat Med 10:51–57

4. Bazzoni G, Dejana E (2004) Endothelial cell to cell junctions: molecular organization and role in vascular homeostasis. Physiol Rev 84:869–901

5. Neuwelt EA, Kroll RA, Pagel MA et al (1996) Increasing volume of distribution to the brain with interstitial infusion: dose, rather than convection, might be the most important factor. Neurosurgery 38:1129–1145

6. Blasberg RG, Patlak C, Fenstermacher JD (1975) Intrathecal chemotherapy: brain tissue profiles after ventriculocisternal perfusion. J Pharmacol Exp Ther 195:73–83

7. Doran SE, Ren XD, Betz AL et al (1995) Gene expression from recombinant viral vectors in the central nervous system after blood–brain barrier disruption. Neurosurgery 36:965–970

8. Bobo RH, Laske DW, Akbasak A et al (1994) Convection-enhanced delivery of macromolecules in the brain. Proc Natl Acad Sci U S A 91:2076–2082

9. Merkus FW, van den Berg MP (2007) Can nasal drug delivery bypass the blood-brain barrier?: questioning the direct transport theory. Drugs R D 8:133–144

10. Illum L (2003) Nasal drug delivery—possibilities, problems and solutions. J Control Release 87:187–198

11. Thorne RG, Pronk GJ, Padmanabhan V et al (2004) Delivery of insulin-like growth factor-I to the rat brain and spinal cord along olfactory and trigeminal pathways following intranasal administration. Neuroscience 127:481–496

12. Hashizume R, Ozawa T, Gryaznov SM et al (2008) New therapeutic approach for brain tumors: intranasal delivery of telomerase inhibitor GRN163. Neuro Oncol 10:112–120

13. Bertram L, Tanzi RE (2005) The genetic epidemiology of neurodegenerative disease. J Clin Invest 115:1449–1457

14. Malhotra M, Nambiar S, Swamy VR et al (2011) siRNA design strategies for effective targeting and gene silencing. Expert Opin Drug Discov 6:269–289

15. Shim MS, Kwon YJ (2010) Efficient and targeted delivery of siRNA in vivo. FEBS J 277:4814–4827

16. Lingor P, Bahr M (2007) Targeting neurological disease with RNAi. Mol Biosyst 3:773–780

17. Gary DJ, Lee H, Sharma R et al (2011) Influence of nano-carrier architecture on in vitro siRNA delivery performance and in vivo biodistribution: polyplexes vs. micelleplexes. ACS Nano 5:3493–3505

18. Perez AP, Mundiña-Weilenmann C, Romero EL et al (2012) Increased brain radioactivity by intranasal ^{32}P-labeled siRNA dendriplexes within in situ-forming mucoadhesive gels. Int J Nanomedicine 7:1373–1385

19. Alvarez-Erviti L, Seow Y, Yin H et al (2011) Delivery of siRNA to the mouse brain by systemic injection of targeted exosomes. Nat Biotechnol 29:341–345

20. Dyer AM, Hinchcliffe M, Watts P et al (2002) Nasal delivery of insulin using novel chitosan based formulations: a comparative study in two animal models between simple chitosan formulations and chitosan nanoparticles. Pharm Res 19:998–1008

21. Vaka SRK, Sammeta SM, Day LB et al (2009) Delivery of nerve growth factor to brain via intranasal administration and enhancement of brain uptake. J Pharm Sci 98:3640–3646

22. Al-Ghananeem AM, Saeed H, Florence R et al (2010) Intranasal drug delivery of didanosine-loaded chitosan nanoparticles for brain targeting; an attractive route against infections caused by AIDS viruses. J Drug Target 18:381–388

23. Ming-Tsung Y, Joan-Hwa Y, Jeng-Leun M (2009) Physicochemical characterization of chitin and chitosan from crab shells. Carbohydr Polym 75:15–21

24. Toan NV (2009) Production of chitin and chitosan from partially autolyzed shrimp shell materials. Open Biomater J 1:21–24

25. Malhotra M, Tomaro-Duchesneau C, Saha S et al (2013) Development and characterization of chitosan-PEG-TAT nanoparticles for the intracellular delivery of siRNA. Int J Nanomedicine 2013:2041–2052

26. Malhotra M, Lane C, Tomaro-Duchesneau C et al (2011) A novel scheme for synthesis of PEG-grafted-chitosan polymer for preparation of nanoparticles and other applications. Int J Nanomedicine 2011:485–494

27. Richard I, Thibault M, De Crescenzo G et al (2013) Ionization behavior of chitosan and chitosan-DNA polyplexes indicate that chitosan has a similar capability to induce a proton-sponge effect as PEI. Biomacromolecules 14:1732–1740

28. Vila A, Sanchez A, Janes K et al (2004) Low molecular weight chitosan nanoparticles as new carriers for nasal vaccine delivery in mice. Eur J Pharm Biopharm 57:123–131

29. Howard KA (2006) RNA interference in vitro and in vivo using a novel chitosan/siRNA nanoparticle system. Mol Ther 14:476–484

30. Alpar HO, Somavarapu S, Atuah AN et al (2005) Biodegradable mucoadhesive particulates for nasal and pulmonary antigen and DNA delivery. Adv Drug Deliv Rev 57:411–430

31. Dong-Won L, Shirley SA, Lockey RF et al (2006) Thiolated chitosan nanoparticles enhance anti-inflammatory effects of intranasally delivered theophylline. Respir Res 7:112

32. Kumar M, Behera AK, Lockey RF et al (2002) Intranasal gene transfer by chitosan–DNA nanospheres protects BALB/c mice against acute respiratory syncytial virus infection. Hum Gene Ther 13:1415–1425

33. Zhu X, Su M, Tang S et al (2012) Synthesis of thiolated chitosan and preparation nanoparticles with sodium alginate for ocular drug delivery. Mol Vis 18:1973–1982

34. Malhotra M, Tomaro-Duchesneau C, Prakash S (2013) Synthesis of TAT peptide tagged PEGylated chitosan nanoparticles for siRNA delivery targeting neurodegenerative diseases. Biomaterials 34:1270–1280

35. Malhotra M, Tomaro-Duchesneau C, Saha S et al (2013) Intranasal siRNA delivery to the brain by TAT/MGF tagged PEGylated chitosan nanoparticles. J Pharm 2013: Article ID:812387, 10 pages. http://www.hindawi.com/journals/jphar/2013/812387/

36. Malhotra M, Tomaro-Duchesneau C, Saha S et al (2013) Systemic siRNA delivery via peptide tagged polymeric nanoparticles, targeting PLK1 gene in a mouse xenograft model of colorectal cancer. Int J Biomater 2013, Article ID: 252531, 13 pages. http://www.hindawi.com/journals/ijbm/2013/252531/

37. Malhotra M, Kulamarva A, Sebak S et al (2009) Ultra-small nanoparticles of low molecular weight chitosan as an efficient delivery system targeting neuronal cells. Drug Dev Ind Pharm 35:719–726

38. Gao X, Tao W, Lu W et al (2006) Lectin-conjugated PEG–PLA nanoparticles: preparation and brain delivery after intranasal administration. Biomaterials 27:3482–3490

39. Wang X, Chi N, Tang X (2008) Preparation of estradiol chitosan nanoparticles for improving nasal absorption and brain targeting. Eur J Pharm Biopharm 70:735–740

40. Gao X, Wu B, Zhang Q et al (2007) Brain delivery of vasoactive intestinal peptide enhanced with the nanoparticles conjugated with wheat germ agglutinin following intranasal administration. J Control Release 121:156–167

41. Gao X, Chen J, Tao W et al (2007) UEA I-bearing nanoparticles for brain delivery following intranasal administration. Int J Pharm 340:207–215

42. Wang J, Chen C, Liu Y et al (2008) Potential neurological lesion after nasal instillation of TiO$_2$ nanoparticles in the anatase and rutile crystal phases. Toxicol Lett 183:72–80

43. Wang J, Liu Y, Jiao F (2008) Time-dependent translocation and potential impairment on central nervous system by intranasally instilled TiO$_2$ nanoparticles. Toxicology 254:82–90

44. Xia H, Gao X, Gua G et al (2011) Low molecular weight protamine-functionalized nanoparticles for drug delivery to the brain after intranasal administration. Biomaterials 32:9888–9898

45. Gao X, Chen J, Chen J et al (2008) Quantum dots bearing lectin-functionalized nanoparticles as a platform for in vivo brain imaging. Bioconjug Chem 19:2189–2195

46. Wen Z, Yan Z, Hu K et al (2011) Odorranalectin-conjugated nanoparticles: preparation, brain delivery and pharmacodynamic study on Parkinson's disease following intranasal administration. J Control Release 151:131–138

47. Zhang Q, Zha L, Zhang Y et al (2006) The brain targeting efficiency following nasally applied MPEG-PLA nanoparticles in rats. J Drug Target 14:281–290

48. Cheng Q, Feng J, Chen J et al (2008) Brain transport of neurotoxin-I with PLA nanoparticles through intranasal administration in rats: a microdialysis study. Biopharm Drug Dispos 29:431–439

49. Alam S, Khan ZI, Mustafa G et al (2012) Development and evaluation of thymoquinone-encapsulated chitosan nanoparticles for nose-to-brain targeting: a pharmacoscintigraphic study. Int J Nanomedicine 2012:5705–5718

50. Md S, Khan RA, Mustafa G et al (2012) Bromocriptine loaded chitosan nanoparticles intended for direct nose to brain delivery: pharmacodynamic, pharmacokinetic and scintigraphy study in mice model. Eur J Pharm Sci 48:393–405

51. Sharma S, Lohan S, Murthy RS (2013) Formulation and characterization of intranasal mucoadhesive nanoparticulates and thermoreversible gel of levodopa for brain delivery. Drug Dev Ind Pharm. doi:10.3109/03639045.2013.789051

52. Ahmad N, Umar S, Ashafaq M et al (2013) A comparative study of PNIPAM nanoparticles of curcumin, demethoxycurcumin, and bisdemethoxycurcumin and their effects on oxidative stress markers in experimental stroke. Protoplasma. doi:10.1007/s00709-013-0516-9

53. Vaka SR, Shivakumar HN, Repka MA et al (2013) Formulation and evaluation of carnosic acid nanoparticulate system for upregulation of neurotrophins in the brain upon intranasal administration. J Drug Target 21:44–53

54. Fazil M, Md S, Haque S, Kumar M et al (2012) Development and evaluation of rivastigmine loaded chitosan nanoparticles for brain targeting. Eur J Pharm Sci 47:6–15

55. Joshi AS, Patel HS, Belgamwar VS (2012) Solid lipid nanoparticles of ondansetron HCl for intranasal delivery: development, optimization and evaluation. J Mater Sci Mater Med 23:2163–2175

56. Wu J, Ding T, Sun J (2013) Neurotoxic potential of iron oxide nanoparticles in the rat brain striatum and hippocampus. Neurotoxicology 34:243–253

57. Liu Z, Jiang M, Kang T et al (2013) Lactoferrin-modified PEG-co-PCL nanoparticles for enhanced brain delivery of NAP peptide following intranasal administration. Biomaterials 34:3870–3881

58. Migliore MM, Vyas TK, Campbell RB et al (2010) Brain delivery of proteins by the intranasal route of administration: a comparison of cationic liposomes versus aqueous solution formulations. J Pharm Sci 99:1745–1761

59. Zhuang X, Xiang X, Grizzle W et al (2011) Treatment of brain inflammatory diseases by delivering exosome encapsulated anti-inflammatory drugs from the nasal region to the brain. Mol Ther 19:1769–1779

60. Eskandari S, Varshosaz J, Minaiyan M et al (2011) Brain delivery of valproic acid via intranasal administration of nanostructured lipid carriers: in vivo pharmacodynamic studies using rat electroshock model. Int J Nanomedicine 6:363–371

Chapter 16

Intrathecal Delivery of Analgesics

Jose De Andres, Juan Marcos Asensio-Samper, and Gustavo Fabregat-Cid

Abstract

Targeted intrathecal (IT) drug delivery systems (IDDS) are an option in algorithms for the treatment of patients with moderate to severe chronic refractory pain when more conservative options fail. This therapy is well established and supported by several publications. It has shown efficacy and is an important tool for the treatment of spasticity, and both cancer and nonmalignant pain. Recent technological advances, new therapeutic applications, reported complications, and the costs as well as maintenance required for this therapy require the need to stay up-to-date about new recommendations that may improve outcomes. This chapter reviews all technological issues regarding IDDS implantation with follow-up, and pharmacological recommendations published during recent years that provide evidence-based decision making process in the management of chronic pain and spasticity in patients.

Key words Intrathecal drug delivery, Opioids, Chronic pain, Baclofen, Spasticity, Ziconotide

1 Introduction

Intrathecal drug delivery is a treatment option in algorithms for treating patients with moderate to severe chronic pain, when other conservative options have failed. This therapy is well established as a valid option effective in the treatment of patients with chronic nonmalignant or malignant pain, and as a tool for management of patients with severe spasticity [1].

The spinal cord is a processing center for pain signals; therefore it is a perfect target for drugs used in management of pain. The advantage of spinal drug delivery lies in the possibility of attaining high drug concentrations at the site of action, with a lower dose requirement than with other routes of administration leading to better symptom control with a lower incidence of side effects, and spinal drug delivery devices enable improved quality of life with increased levels of patient satisfaction [1].

The disadvantages of these systems, relating to technical implantation issues, pharmacologic reactions or possible side

Kewal K. Jain (ed.), *Drug Delivery System*, Methods in Molecular Biology, vol. 1141,
DOI 10.1007/978-1-4939-0363-4_16, © Springer Science+Business Media New York 2014

effects following spinal drug delivery, as well as human error when programming or filling the storage device, are associated with a high risk of morbidity and mortality, making this a type of therapy that should be managed at specific centers. All these issues, together with the financial and human cost of setting up and maintaining this type of therapy, make it essential to keep up to date with new aspects of management in order to optimize outcomes [2].

In this respect, to take over from guidelines issued in 2007, the 2012 Polyanalgesic Consensus Conference (PACC 2012) Consensus Documents were published in 2012, with the aim of providing indications, standards of use, and updated management for intrathecal drug infusion devices [1].

2   Selection and Indications for Intrathecal Therapy in Chronic Pain

The true incidence of patients requiring interventional analgesic techniques remains unknown because of varying inclusion criteria and practices in different centers. Implantable IT drug delivery systems (IDDS) have been used in various chronic painful conditions for more than 30 years.

2.1   Chronic Malignant Pain

Intrathecal drugs (opiates and non-opiates with or without adjuvants) can be an effective treatment option in a select group of patients with chronic cancer pain refractory to conventional medical treatment [3].

A review of current published evidence concerning intrathecal drug infusion in patients with chronic malignant pain provides updates on indications for using this type of therapy, precautions, contraindications, and possible side effects, including also socioeconomic issues related to its use and management in difficult or uncommon situations [4]. The authors identify some important features of cancer patients, including the need to use drug combinations to achieve adequate analgesia because a mixed (nociceptive/neuropathic) pain type is common, the need to evaluate past medical history in order to avoid potential contraindications or drug interactions that might compromise patient safety, and the required pre-implantation assessment of the patient's spinal anatomy, as well as the existence of primary or metastatic tumor pathology, and associated nonmalignant diseases such as osteoarthritis, arthritis, scoliosis, spasticity, in order to prevent complications as far as possible.

Along the same lines, PACC 2012 [5] notes other features typical of cancer patients, recognizing that, because time is critical in this type of patient, the pre-implant test may become unnecessary or even counterproductive. It also criticizes selection criteria given in other consensus documents based on cancer patients' life expectancy (in which this therapy is considered more cost-effective than medical treatment in patients with a life expectancy of over 6 months), recognizing that life expectancy can improve in patients treated with intrathecal infusion as side effects are reduced and quality of life is improved globally irrespective of initially predicted life expectancy.

PACC recognizes that the cancer patient is a complex patient who should be assessed at a tertiary center, and in whom the standard recommendations are sometimes invalid, as drugs are used at such high doses or concentrations or in combinations that would be deemed "unacceptable" based on the guidelines.

2.2 Chronic Nonmalignant Pain

Chronic nonmalignant pain has been defined as the type of pain that lasts a long time, generally more than 6 months. It can be due to various kinds of noncancer causes, particularly degenerative and inflammatory processes and complex regional pain [6]. Intrathecal drug delivery in these cases has limited levels recommendation in the literature [2]. One of the most important studies published to date 20 patients with chronic nonmalignant pain treated with intrathecal drugs were monitored for an average of 13 years, which is one of the longest follow-up periods reported in the literature on patients fitted with an intrathecal drug system [7]. Statistically significant improvement was observed for sensory and psychosocial variables: pain intensity, pain relief, depression/anxiety, quality of life, self-sufficiency, mobility, sleep, and social life. The authors concluded that intrathecal drug infusion is an appropriate solution in the long-term management of selected patients with chronic nonmalignant pain.

Other authors, however, see inconsistency in the use of intrathecal drug infusion in patients with chronic nonmalignant pain, and question the risk–benefit ratio in these patients [8]. Their clinical impression is that this therapy provides moderate analgesic benefit in the first 6 months post-implantation, and that the benefit declines over time. They also report loss of the patient's overall functional capacity, reinforcement of passive disease-coping behaviors, and reinforcement of the patient's illness role. On this basic assumption, they describe ceasing intrathecal opioid therapy and switching to oral/transdermal administration in a cohort of 25 patients. Disadvantages associated with treatment cessation included: transient withdrawal symptoms, increased pain, and reduced physical activity. In contrast, patient-observed advantages following cessation of opioid infusion included reduced side effects (sweating, weight gain, edema), withdrawal of testosterone replacement therapy in some cases, improved comfort due to disappearance of the abdominal mass effect caused by the infusion device, and less hospital dependence because of fewer visits to the pain unit for treatment follow-up.

With these findings, and acknowledging that there is no evidence from randomized clinical trials concerning the use of intrathecal infusion in chronic nonmalignant pain, or about the onset of tolerance with prolonged use of systemic opioids, they conclude that ceasing established intrathecal therapy in patients with chronic nonmalignant pain may be an appropriate management option. A safe method of abruptly stopping spinal opioid infusion using standardized protocols has been described by using buprenorphine and clonidine [9].

2.3 Patient Selection and Common Indications

The placement of an intrathecal catheter, tunneling of the catheter, pocketing for pump placement, and connection of the system are a complex process of interventions requiring great technical skill. The selection of the patient who will receive the pump is as important as each step in the procedural process. Selection considers patient characteristics, while indications consider the disease state being treated. In general, selection criteria for patients eligible for implantation of an intrathecal drug infusion pump are as follows [10, 11]:

1. Patient with moderate to severe pain (VAS > 4), especially multifocal.

2. Failure of conservative treatment measures.

3. Test phase provides adequate pain control (>50 % improvement for at least 10 h is traditionally considered adequate), with tolerable side effects and functional improvement.

4. The patient has a poor treatment response and suffers unacceptable side effects with oral/transdermal drug use.

5. The patient has good spinal anatomy for implantation of a spinal infusion system.

6. The patient understands the therapy, has realistic treatment expectations, and knows the possible side effects. Informed consent.

7. The patient has no contraindication to implantation because of chronic hematologic conditions (severe coagulation disorders, aplastic anemia, etc.).

8. The patient has no contraindication to implantation because of an active infection.

9. The patient has no skin problems to suggest a post-implantation foreign-body reaction, and no past history of allergy to drugs or typical infusion system components.

10. The patient has no psychiatric/psychologic abnormalities to contraindicate implantation, and no history of substance abuse (alcohol, drugs).

11. The implant team is trained not only in implanting the system but also in its subsequent management (dose escalation, refills, resolving complications, etc.).

Possible selection indications for intrathecal infusion pump implantation are [10]:

1. Chronic malignant pain: primary malignancies or metastatic lesions causing pain from tissue invasion, neuropathic pain secondary to chemotherapy treatments or radiotherapy.

2. Chronic noncancer pain: failed spinal surgery syndrome, spinal canal stenosis, spondylolisthesis, vertebral fractures, peripheral neuropathies, severe osteoarthritis, complex regional pain syndrome, connective tissue diseases, etc.

3 Drugs Used in Intrathecal Therapy

The selection of the proper drug to use in an individual patient is complicated by disease state, patient characteristics, and the character of the pain which afflicts the patient. In selecting the proper drug for the patient, the physician should attempt to determine the type of pain. Neuropathic pain syndromes respond less frequently to solo opioid therapies, as opposed to pure nociceptive pain syndromes, which respond well to opioids. The occurrence of side effects may also have a profound effect on making adjustments to infusion combinations [5].

Since the publication of the third PACC report in 2007, the published literature on IT therapy has expanded, and the 2012 updated version of the PACC treatment algorithm is based on the best available evidence from published reports and on panel discussions. The PACC 2012 algorithm for drug selection provides clinical practice guidelines for the optimization of IT therapy with single-drug treatments and drug combinations in a rational and prioritized order [5]. Unlike the previous PACC algorithms, the current algorithm contains separate arms for neuropathic, nociceptive, and mixed pain states. This is important as the literature suggests that the efficacy of this therapy differs in patients suffering from one type of pain or another [12]. In each arm of the algorithm, medications are arranged in a hierarchy on the basis of evidence of efficacy and safety (*see* Tables 1, 2, and 3). First-line medications/combinations are supported by extensive clinical experience and published clinical and preclinical literature and are typically used to initiate IT therapy. Notably, morphine and ziconotide are the only two agents approved by the US Food and Drug Administration (FDA) for IT analgesia, while morphine and ziconotide are approved for use in Europe, the Middle East, and Africa. The use of other agents is common among pain practitioners who manage IT pumps, and discussions of such agents are included to give guidance on safe practice and patient safety. These algorithms were created to help guide clinicians in the safe and effective use of IT therapy; however, physicians should use their own best clinical judgment in making treatment decisions for their patients [5].

Table 1
Intrathecal (IT) therapies in neuropathic pain

Line1	Morphine	Ziconotide	Morphine + bupivacaine
Line2	Hydromorphone	Hydromorphone + bupivacaine or hydromorphone + clonidine	Morphine + clonidine
Line3	Clonidine	Ziconotide + opioid	Fentanyl
Line4	Opioid + clonidine + bupivacaine	Bupivacaine + clonidine	
Line5	Baclofen		Fentanyl + bupivacaine or fentanyl + clonidine

Line1: Morphine and ziconotide are approved by the US Food and Drug Administration for IT therapy and are recommended as first-line therapy for neuropathic pain. The combination of morphine and bupivacaine is recommended for neuropathic pain on the basis of clinical use and apparent safety. *Line2*: Hydromorphone, alone or in combination with bupivacaine or clonidine, is recommended. Alternatively, the combination of morphine and clonidine may be used. *Line3*: Third-line recommendations for neuropathic pain include clonidine, ziconotide plus an opioid, and fentanyl alone or in combination with bupivacaine or clonidine. *Line4*: The combination of bupivacaine and clonidine (with or without an opioid drug) is recommended. *Line5*: Baclofen is recommended on the basis of safety, although reports of efficacy are limited

According Polyanalgesic Consensus Conference 2012 [5]

Table 2
Intrathecal (IT) therapies in nociceptive pain

	Morphine	Hydromorphone	Ziconotide	Fentanyl
Line1	Morphine	Hydromorphone	Ziconotide	Fentanyl
Line2	Morphine + bupivacaine	Ziconotide + opioid	Hydromorphone + bupivacaine	Fentanyl + bupivacaine
Line3	Opioid (morphine, hydromorphone, or fentanyl) + clonidine			
Line4	Opioid + clonidine + bupivacaine	Bupivacaine + clonidine		
Line5	Sufentanil + bupivacaine + clonidine			

Line1: Morphine and ziconotide are approved by the US Food and Drug Administration for IT therapy and are recommended as first-line therapy for nociceptive pain. Hydromorphone is recommended on the basis of widespread clinical use and apparent safety. Fentanyl has been upgraded to first-line use by the consensus conference. *Line2*: Bupivacaine in combination with morphine, hydromorphone, or fentanyl is recommended. Alternatively, the combination of ziconotide and an opioid drug can be employed. *Line3*: Recommendations include clonidine plus an opioid (i.e., morphine, hydromorphone, or fentanyl) or sufentanil monotherapy. *Line4*: The triple combination of an opioid, clonidine, and bupivacaine is recommended. An alternate recommendation is sufentanil in combination with either bupivacaine or clonidine. *Line5*: The triple combination of sufentanil, bupivacaine, and clonidine is suggested

According Polyanalgesic Consensus Conference 2012 [5]

Table 3
Recommended starting dosage ranges of intrathecal medications

Drug	Recommended starting dosage
Morphine	0.1–0.5 mg/day
Hydromorphone	0.02–0.5 mg/day
Ziconotide	0.5–2.4 µg/day
Fentanyl	25–75 µg/day
Bupivacaine	1–4 mg/day
Clonidine	40–100 µg/day
Sufentanil	10–20 µg/day

According Polyanalgesic Consensus Conference 2012 [5]

3.1 Opioids

Morphine. First-line treatment in patients with chronic neuropathic and/or nociceptive pain, including pain of nonmalignant and malignant origin [5, 13].

Hydromorphone. First-line treatment in patients with chronic nociceptive pain and second-line treatment as monotherapy or combined with bupivacaine or clonidine in patients with chronic neuropathic pain, the choice being constrained by price (hydromorphone is approximately 30 times more expensive than morphine in the European market) [14].

Fentanyl. Alone or combined with bupivacaine or clonidine, this is regarded as third-line treatment in cases of chronic neuropathic pain. As a new feature in cases of nociceptive pain, fentanyl is now considered first-line treatment because of its good profile in cases of prolonged infusion (lower incidence of granuloma onset).

Sufentanil, Methadone, Meperidine. No relevant new studies have appeared concerning the use of these drugs in spinal infusion.

3.2 Non-opioids

Ziconotide. This is the first-line drug in chronic nociceptive and neuropathic pain [5, 15, 57]. The PACC 2012 consensus document reviews the literature on ziconotide use, including the latest data on its long-term efficacy as monotherapy or combined with other drugs, its safety, and its use in the test phase [5, 16]. Successful use of combined ziconotide and morphine is recognized in patients with chronic cancer pain refractory to systemic treatment [17].

The advantages of using ziconotide include its morphine-independent mechanism of action, absence of respiratory depression, and the low doses needed to achieve clinical effects. However, ziconotide use poses a challenge because of the high number of side effects, many of which are extremely serious. An observational study showed that low initial doses of ziconotide followed by slow dose titration reduces the incidence of serious side effects related to its use in patients with chronic cancer pain, but will not affect the incidence of onset of mild to moderate adverse effects [18].

Baclofen. This γ-aminobutyric acid (GABA) agonist is a fifth-line treatment in chronic neuropathic pain [5]. Baclofen is also used in spinal injury, brain damage, spasticity, amyotrophic lateral sclerosis, cerebral palsy, and stiff-man syndrome. Use of baclofen in severe, progressive spasticity that is refractory to conventional medical treatment is considered a good treatment option [19, 20]. It is useful in children with progressive neurologic disease [21], in patients with dystonia [22, 23], myoclonus [24], dysautonomia and hypertonia following severe head injury [25]. In patients with complex regional pain syndrome with associated dystonia, significant improvement in overall pain management has been demonstrated a during the first 6 months, but analgesic efficacy decreases after that despite clinical improvement in dystonia, which requires increase of intrathecal baclofen dose [26].

Possible adverse effects include a mean weight gain of 5.43 kg in children on IT baclofen (ITB) for longer than 1 year [27]. There may be a rise in respiratory adverse events (increased "respiratory disturbance index" and apnea of central origin) in patients on ITB, especially if delivered in "bolus" mode [28]. Paradoxical worsening of spasticity and pain has been reported in a patient on spinal baclofen treatment coinciding with baclofen dose escalation that was considered to be probably due to saturation and downregulation of the $GABA_B$ receptor [29]. Although uncommon, the possibility of infection during implantation of the intrathecal device must also be borne in mind, which can lead to removal of the device in 59 % cases [30].

4 Considerations Prior to Implanting an Intrathecal Pump

4.1 Pre-implant Trialing

Traditionally, IT trials have been used to determine patient response to therapy and to establish a baseline measurement from which potential improvement can be assessed, and were also considered mandatory before device implantation, yet no standard method of pre-implant trialing has been established and the locations in which testing is to be performed vary greatly. Similarly, ideal trial duration has not been confirmed. Trialing can be initiated through a single injection, multiple injections, or continuous infusion, either in the intrathecal or in the epidural space. The decision to use one method over another is largely based on the physician's preference, availability of facilities, practice environment, and health insurance coverage provided [4].

A trial is considered positive when there is more than 50 % reduction in pain intensity measured with the available tools, indicative of further success with IT therapy. Whereas trials may play an important role in long-term therapy, they have limited value for patients receiving palliative care where it is essential to be expeditious in evaluating treatment options. Additionally, a direct correlation between a patient's response during an IT trial and subsequent effects of therapy has yet to be established, thus limiting the predictive value of this screening technique [4].

4.2 Psychologic Assessment

There are various reasons for suggesting that patients eligible for an implantable neuromodulation system need to have their psychologic profile assessed including the multidimensional component of pain consisting both of sensory and of emotional and cognitive dimensions in particular [31]. The need for personal adjustment required by invasive techniques of this type has also been emphasized. This adjustment is necessary for therapeutic success, defined by the meeting of expectations, and in terms of the criteria by which candidate patients for these systems define their quality of life. Lastly, treatment adherence is undeniably necessary for maintaining the system once implanted, which is expensive for health services.

In this respect, the European Federation of Chapters of the International Association for the Study of Pain (EFIC) suggested in 1998 that assessment and screening of patients eligible for neuromodulation systems should take place at specialist pain treatment centers in the context of a multidisciplinary approach, patients should be thoroughly informed about the treatment, and in-depth psychologic assessment should be done at the start of treatment screening [32].

The process of evaluating candidate patients for neuromodulation system implantation should include assessment of their psychologic profile. In general, assessment should follow three basic criteria: (1) performance of a clinical interview, (2) review of medical records, and (3) psychometric assessment. For psychometric assessment, it is advisable to select measurement instruments that are necessary and sufficient for the end purpose. The use of these instruments will depend on the individual case and the variables to be confirmed. It should fit the context of the medical records review and clinical interview. Psychometric test results cannot be used as the sole assessment criterion. The instruments selected will depend on professional judgment according to the individual case [33]. Those used have included:

5   Personality and Psychopathologic Profile

- Minnesota Multiphasic Personality Inventory 2 (*MMPI-2*) (Hathaway SR and McKinley JC, 1989). Spanish adaptation [34].
- Millon Clinical Multiaxial Inventory (*MILLON-II*) (Millon T, 1997). Spanish adaptation [35].

6   Coping

- Pain Coping Questionnaire (reduced version) (CAD-R) [36].
- Coping Strategies Questionnaire (*CSQ*) (Rosenstiel and Keefe, 1983), Spanish adaptation [37].

7   Anxiety and Depression

- Hospital Anxiety and Depression Scale (*HAD*) (Zimong AS and Snaith RP, 1983), Spanish adaptation [38].
- State-Trait Anxiety Inventory (*STAI*) (Spielberger CD, Gorsuch RL, and Lushene RE, 1968), Spanish adaptation [39].
- Beck Depression Inventory (*BDI*) (Beck AT, 1979), Spanish adaptation [39].

8   Pain Assessment and Beliefs

- Pain Beliefs and Perceptions Inventory (*PBAPI*) [41].
- Visual Analogue Scale (*VAS*) [42], McGill Pain Questionnaire, Spanish Version (*MPQ-SV*) [43], Lattinen Scale [44].

Following psychologic assessment there are three possible situations:

1. *Unsuitable*: This decision is made when any variable distorting the possible therapeutic benefit is considered insoluble. This conclusion should be reached when it is absolutely certain that these distorting variables cannot be reversed by any type of intervention.

2. *Temporarily unsuitable*: This decision is made when it is thought that the patient's profile may be changed or affected by intervention, by either the medical or the psychologic team. The percentage of patients who may benefit from this situation is about 35 % of those initially assessed.

3. *Suitable for implantation*: This decision is based not only on analysis of the biomedical variables that may be involved in the patient's condition, but also on multidisciplinary assessment of the patient, and involving him or her in decision-making.

8.1 Cost-Effectiveness Study

Cost–benefit analysis for a therapy should be conducted individually for each patient while simultaneously taking an overview of the care plan within the patient's healthcare process for his or her particular problem. This is the only possible objective approach to intrathecal drug therapy for the treatment of chronic pain and intractable spasticity. Therapies of this type require a large initial financial investment, although they are cost-effective in the medium term, especially if assessed globally in comparison to other, noninvasive therapies. It is also evident that optimization of intrathecal therapy is achieved when indication, implantation and follow-up are performed by the application of management algorithms at centers specializing in a multidisciplinary approach to patients with chronic pain and intractable severe spasticity.

Selecting patients properly, minimizing complications, and maintaining long-term efficacy are key to success of the therapy, so this is the way to rationalize its high initial cost. With regard to the first point, candidate patients for intrathecal drug infusion must show evidence of disease, failure of conservative treatments, response to preliminary trials, no neuropsychiatric abnormality or allergy to the drugs used, and must have a life expectancy in excess of 3 months.

Under these conditions, intrathecal drug infusion for the treatment of chronic pain and spasticity provides improved patient quality of life and a reduction in hospital stays with Evidence Level A, according to a 1999 analysis by the Health Quality Agency [45].

The ability of intrathecal therapy to prove cost-effective despite the large initial implantation cost can be explained by superior long-term analgesic efficacy, which enables the patient to reduce the number of routine and emergency consultations. The lower incidence of adverse effects arising from reduction in the effective dose of the drug means not only less use of health services but also lower consumption of concomitant medication. Increased patient quality of life reduces carer workload, and even enables 20 % of patients to undergo occupational rehabilitation at 12 months of follow-up. These patients also experience a considerable reduction in consumption of analgesic and adjuvant medication by the oral and other administration routes, of up to 65 % at 6 months of treatment. Lastly, the high degree of patient satisfaction with the therapy and the general care plan offered by institutions, which reaches 96 % in some series, likewise results in better use of health services and consequent long-term savings [46].

Several clinical trials have demonstrated superior efficacy of intrathecal therapy for pain symptom control with few side effects compared with conventional multimodal therapy. However, in one study 60 % improvement in 6-month survival was demonstrated in the cancer patient group treated by intrathecal therapy for pain relief [47]. In this respect, the potential increase in survival derived from pain relief means better coping with the costs of therapy, which becomes more cost-effective as treatment time increases. Although it has been established that the minimum life expectancy of a candidate patient for intrathecal therapy is 3 months, it could be said that personalized case analysis, and comparative study of the therapy versus other therapies with a dubious effect on quality of life or survival, might lead it to be indicated in patients with less than 3 months life expectancy. In this respect, bisphosphonate therapy for bone metastases, the use of new-generation antiemetics, or palliative chemotherapy can entail very high costs in patients of this type, particularly in inpatient applications. A 3-month period of intrathecal therapy makes it cost-effective, not just versus noninvasive therapy and nonanalgesic therapy in cancer patients, but also versus analgesia via external epidural catheters and patient-controlled analgesia in the outpatient setting.

In patients with chronic noncancer pain, evidence suggests that intrathecal therapy is a cost-effective method when it reaches 22 months of treatment. Even after 5 years of treatment, it still has lower annual and cumulative costs than conventional multimodal therapy. Therefore, in patients with intractable chronic noncancer pain too, because of its optimum efficacy and adverse effects profile, intrathecal therapy has a good cost-effectiveness ratio, especially when based on proper patient selection and implantation technique. Screening patients by means of a test period of continuous intrathecal infusion, or even by the epidural route for morphine treatment, has proved very helpful for evaluating pharmacologic response and realistic patient treatment expectations, prior to definitive treatment [46].

It is perhaps in the treatment of intractable severe spasticity that intrathecal baclofen therapy offers its best cost–benefit profile, possibly because of its high efficacy and safety in patients in whom conventional therapy consumes a lot of healthcare resources in terms of personnel and hospitalization. This is why some protocols recommend prompt indication in multiple sclerosis patients with an Expanded Disability Status Scale of 7 or above with Evidence Level A, and over 5 with Evidence Level C. These guidelines also stress the need to attend centers specializing in these therapies for assessment, implementation, and spasticity management [46].

Likewise for intrathecal baclofen therapy, it is advisable to use objective assessment scales previously and to check patient progress in order to monitor treatment response, as well as measurement scales for quality of life, survival, functional activities of daily living, and general state of health. These scales are very useful for objectively examining the patient's prior situation, demonstrating the need for this therapy rather than others, and assessing the subsequent response and its consistency over time.

9   Complications of Intrathecal Drug Delivery

Although IT opioid infusion is an innovative approach to the treatment of chronic severe pain, the risks and complications of this therapeutic modality are now beginning to be more thoroughly recognized. Morbidity and mortality should also be considered.

Complications of IDDS can be divided into the following categories [4]: (1) surgical complications, (2) mechanical complications, (3) pharmacologic, (4) medical, and (5) catheter tip granuloma.

9.1   Surgical Complications

Bleeding. Bleeding can occur from ineffective hemostasis, preoperative anticoagulation, vascular injury, and secondary hemorrhage. Bleeding in deep intraspinal or epidural space, while extremely rare, is associated with increased neurologic morbidity. Fluoroscopy is essential to avoid periosteal or rare spinal epidural tumor trauma. Significant bleeding in epidural space may lead to epidural hematoma, spinal cord compression, and paraplegia. It will present as increasing backache and neurologic deficit, and emergency neurosurgical intervention for decompression might be required to prevent permanent neurologic injury. Superficial postoperative bleeding or hematoma at the pump site may present as swelling, pressure, and pain. Leakage of serosanguineous fluid may or may not be present from the wound itself. Often the site will have diffuse bruising. This is a self-limiting problem that requires monitoring. Abdominal binders are sometimes helpful by applying direct pressure over the site to reduce swelling and discomfort.

Infection. Infection is one of the preventable complications; use of strict sterile techniques, proper antibiotic, and frequent monitoring can prevent serious consequences. Some practitioners advocate the use of intraoperative antibiotic irrigation as well, with frequent monitoring of the implant site for increased pain, erythema, tenderness, swelling, drainage, fever, and leukocytosis. Removal of the implant is not necessary in most cases. Superficial infection should be cultured and treated with proper antibiotics. More serious infection involving the pocket or catheter-related infection will require removal of the device. Infection involving epidural or intrathecal spaces requires immediate removal of all implant devices and administration of intravenous antibiotics. Intrathecal infections are rare and present with fever, nuchal rigidity, and leukocytosis. Cerebrospinal fluid (CSF) should be sent for culture in suspected infection. Epidural abscess may require urgent computed tomography (CT)/magnetic resonance imaging (MRI) and prompt neurosurgical or spinal intervention to prevent severe neurologic complication.

Leakage of cerebrospinal fluid. Persistent CSF leak is a known complication of IT pump system insertion for drug delivery and can occur in as many as 20 % of patients with an implant in place. The CSF leak may occur due to either dural tear or misconnection or improper connection or leakage from the catheter. Persistent CSF leakage can lead to all complications of postdural puncture headache (PDPH) and its sequelae. Catheter misconnection and leakage should be ruled out in patients with severe PDPH symptoms; misconnection and leakage can be diagnosed by failure to aspirate CSF from the pump port or collection of CSF around the pump. Pericatheter CSF leakage can be managed conservatively with increased fluid intake, simple analgesic, bed rest, caffeine, etc. Severe symptoms may require epidural blood patch, surgical closure of dural tear, repositioning of catheter, and purse-string sutures over the dura around the catheter. In severe leakage, a hygroma may develop, which is an accumulation of CSF subcutaneously near the dorsal incision. Aspiration of this fluid should be avoided due to the risk of infection. A large leak draining from the incision may require surgical intervention.

Seroma. Wound seroma around the pump pocket or in the back wound has been observed at the site of catheter insertion. Back wound seroma may be due to persistent pericatheter CSF leakage. Pump pocket seroma may develop in patients with hypoalbuminemia and in patients with venous or lymphatic obstruction. It might also develop as an aseptic foreign-body reaction. This type of seroma can last for 1–2 months and is usually self-limiting or requires a simple abdominal binder. In problematic and painful seromas, placement of a drain and use of an abdominal binder should be considered. If infection is suspected, fluid should be sent for culture and appropriate systemic antibiotic therapy should be initiated. Tetracycline or doxycycline 1–2 g diluted in 20 cc of saline may be injected into the pocket after seroma drainage to seal persistent fluid accumulation (as in pleurodesis). Even with gross ascites and aberrant venous drainage, seromas frequently resolve spontaneously. In the absence of resolution, movement or removal of the pump should be considered.

9.2 Mechanical Complications

Catheter pump misconnection. A catheter pump misconnection typically occurs immediately following implantation but presentation may be delayed.

Loss of pump propellant. The loss of pump propellant can be revealed as an altered (excessive or reduced) rate of drug delivery and may result in a variety of symptoms, including overdose and acute withdrawal adverse effects.

Gear shaft wear and motor stall. These malfunctions lead to symptoms of drug underinfusion and may not be accompanied by an alarm.

Leakage of administered agent. Leakage may occur at the catheter-pump connection during the postoperative period or can be delayed in onset; these malfunctions can have several causes, including a needle piercing the catheter wall during infiltration of an additional local anesthetic utilized during incision-site closure, trauma to the catheter caused by self-retaining or handheld "cat's paw" retractors, or catheter kinking proximate to the pump.

Intrathecal catheter displacement. Displacement of the catheter may result in CSF leakage, which can cause local hygroma.

Intrathecal catheter kinking. Kinking of the IT catheter can occur at any location, from the pump to the catheter receiving device; kinking makes it difficult or impossible to aspirate CSF and inject agents into the pump port site.

9.3 Pharmacologic

Medication errors are preventable complications. The manufacturer's instruction manual should be strictly followed. Drug refill programming must be done by trained personnel who can actually assess pain accurately, conduct physical examinations, and assess subtle changes in condition. Strict aseptic technique should be applied during each refill; the concentration and combination of drugs should be used as per the guideline. Extreme vigilance must be given to all aspects of safety, particularly the prevention of inadvertent administration of drugs by the wrong route. Most medication complications are either due to hypersensitivity or of the allergic type. These symptoms are often minimal and can be diminished or eliminated by slow titration of medications. Patient education is key to successful IT analgesic therapy. Patients should be constantly reminded of the active role they have to play in their therapy, as well as the restrictions associated with the implanted device, and to report to the medical team in case of any change in their medical condition.

9.4 Medical

A retrospective study reports decreased libido of up to 96 % in men and 69 % in women. In a study of 73 patients with noncancer pain, the majority of the patients developed hypogonadotropic hypogonadism; 15 % developed central hypocortisolism. Hormone replacement ameliorated these effects.

Intrathecal opioid. Morphine is considered the gold standard because of its stability, receptor affinity, and extensive experience of using the drug by this route. When there is intolerance to morphine, hydromorphone can be used. This is five times more potent than morphine with a similar side effects profile. Centrally mediated side effects of IT opioids include late respiratory depression, pruritus, nausea, vomiting, urinary retention, sedation, constipation, edema, weight gain, excessive sweating, memory or mood changes, and headache.

Intrathecal local anesthetics. Intrathecal bupivacaine has been used in combination with morphine for better pain control in both malignant and nonmalignant pain. Intrathecal bupivacaine can be used to control neuropathic pain. There is evidence that bupivacaine acts synergistically with morphine, reducing the need for increased IT morphine doses. Local anesthetics can cause sensory deficits, motor impairment, signs of autonomic dysfunction, and neurotoxicity. This is less likely to be a problem if continuous infusions rather than boluses are used. Clinically relevant side effects are not usually seen at bupivacaine doses of less than 15 mg/day. At higher doses, urinary retention, weakness, fatigue, somnolence, and paresthesia have been observed.

Intrathecal clonidine. This is generally used in combination with morphine and/or bupivacaine. The admixture of clonidine and morphine acting synergistically has been shown to be effective in patients with cancer pain and spinal cord injury. The most common side effects of IT clonidine are hypotension, bradycardia, and sedation.

Intrathecal ziconotide. Ziconotide produces its analgesic effects by blocking specific N-type calcium channels found at presynaptic terminals in the dorsal horn. Ziconotide can be initiated at 2.4 mg/day and titrated according to analgesic response and adverse effects. The maximum dose is 21.6 mg/day. Mixtures of ziconotide with other IT medications including morphine, hydromorphone, clonidine, and baclofen are associated with a reduction in ziconotide concentration of the order of 20 % within a few weeks. Side effects with ziconotide include dizziness, nausea, nystagmus, gait imbalance, confusion, and urine retention. Serious but rare side effects include psychosis, suicide, and rhabdomyolysis.

9.5 Catheter Tip Granuloma

Development of an inflammatory mass at a catheter tip is a rare but potentially devastating complication of long-term IT administration of analgesic medications [48–50]. The reported incidence of IT granuloma is 0.04 % after 1 year of therapy and 1.16 % after 6 years of therapy. There are more than 500 reported cases worldwide. The incidence is estimated at 0.5 % per patient per year. These granulomas are found between the spinal cord and the dura and occur mostly in the thoracic area. They can cause spinal cord compression, affecting motor and sensory function, and radicular pain in thoracic or lumbar regions. There is failure of analgesia as drugs are unable to reach target neural tissue. The etiology is unknown but may be a reaction to the catheter tip, or a low-grade infection, or possibly a reaction to infused medication. Several factors contribute to the development of IT granuloma, including the agent infused, catheter position, low CSF volume, and drug dose and concentration. Although IT granuloma is more common with morphine it can occur with any medication, such as sufentanil, baclofen, and clonidine; its onset appears to be within several months after utilization of IT therapy. Typically, individuals present with an increase in pain intensity that precedes signs and symptoms of neurologic deterioration. Practitioners should suspect development of a granuloma if new-onset pain appears, or increasing pain worsens despite escalating doses of IT opioid, or neurologic symptoms arise. MRI remains the gold standard for surveillance when evaluating the presence of a catheter-related inflammatory mass, although CT/myelogram through the pump offers a more cost-effective technique. Failure to diagnose this condition could lead to permanent neurologic injury. The treatment of IT granuloma is determined by the clinical condition of the patient. If there are no neurologic deficits, then pump medication should be replaced with sterile normal saline. Monthly serial MRI to observe the regression of the mass is recommended. Once symptoms resolve, restart a non-offending opioid and monitor the patient closely for recurrence by MRI every 3 months. If neurologic deficits are present, then surgical intervention and removal of the IT catheter is indicated. The pump can be kept in place if another catheter placement is deemed appropriate. There is direct correlation between drug concentration and IT granuloma, so the IT drug should not exceed the recommended concentration.

10   Intraspinal Drug Delivery Systems

Intraspinal drug delivery systems as used in Spain, but as well internationally, have been reviewed elsewhere [51].

10.1   External Systems

Under sterile conditions, an intraspinal catheter is inserted, usually at the lumbar level, and fixed directly to the skin or tunneled subcutaneously a few centimeters (5–10 cm) beyond the insertion site. The catheter is then connected to an antibacterial filter. Medication is delivered continuously via an ambulatory infusion pump.

This has a high infection rate and therefore a greater meningitis risk, which increases with catheter dwell time. Whenever it is used for more than a week, tunneling of the catheter is recommended.

10.2   Partially External Systems

This type of system is mainly used for the test phase of intraspinal drug treatment, prior to definitive implantation. In this system, the distal end of the intrathecal catheter is tunneled to the anterior wall of the abdomen, where it is connected to a reservoir positioned in the subcutaneous tissue.

The reservoir is accessed by the percutaneous route with a right-angled needle, with a special tip that prevents damage to the silicone seal and does not carry microparticles of silicone into the system. Medication can be delivered in boluses or, more commonly, to avoid handling the system and for greater patient convenience, via an ambulatory infusion pump, which can be programmed with a background continuous infusion and boluses on patient demand.

Radiopaque silicone or polyurethane catheters are used, which are more durable than normal catheters, with fewer kinking or breakage complications.

These systems have a lower risk of infection than external systems. They are easy to manage and cheap, but have the disadvantage that the patient has to wear a portable infusion pump, which hinders patient hygiene.

In general, their use is restricted to cancer pain in patients with a life expectancy of less than 3 months. Some groups use these systems for long test phases (lasting several weeks) prior to definitive implantation.

10.3   Implanted Systems

Implantable devices can provide effective therapies but are invasive as well as expensive, and should be used only when more conservative and less costly therapies have failed to provide relief of pain and suffering [52]. Their high cost restricts the use of these to patients with life expectancy of more than 3–6 months, which means they are most widely employed in the treatment of chronic nonmalignant pain and spasticity.

In these systems, both the catheter and the infusion system are completely implanted, which reduces the probability of the system becoming infected and gives the patient great independence. The catheters are made of a radiopaque silicone or polyurethane material. Theoretically, segmental analgesia can be obtained if the catheter tip is left close to the pain-producing area [51]. Positioning the catheter tip away from the pain-producing area reduces the efficacy of lipid-soluble drugs, such as local anesthetics, fentanyl, etc.

This category of implanted systems consists of two types: fixed-flow systems and programmable-flow systems.

Continuous-flow systems work by propulsion generated either by a gas contained in a chamber that exerts constant pressure on the medication reservoir or by a diaphragm made of a polymer that performs the same function.

The advantage of these systems is that they are much cheaper than programmable systems. Also, as they are not powered by a battery, they do not have to be replaced. The biggest disadvantage is the requirement to refill the pump every time the dose delivered to the patient needs changing. The time between refills depends on the rate of flow (ml/day) and the volume of the medication chamber.

Programmable-flow systems allow greater adjustment to variations in the patient's analgesic needs over time. These systems, powered by lithium batteries, contain a processor that controls the pump flow rate, which is programmed externally by telemetry. It is thus possible to alter the infusion modality and daily dose without having to insert a needle in the patient. These systems in turn enable different infusion modalities to be programmed according to the needs of each patient (continuous infusion, boluses at preset times, infusion with preset boluses, variable infusion with increased or decreased flow depending on the time of day for adjustment to patient needs). The system has to be replaced once the life of the lithium battery has expired (5–9 years). Personalized therapy is possible by enabling the patient to use an external control to deliver an extra bolus dose of preset characteristics.

11 Implantation Technique

The decision to insert a pump is a complex medical problem that requires careful evaluation, proper planning, and technical skills. Before implanting an infusion pump, as discussed above, an intrathecal efficacy test of the chosen drug should be performed. The system must be implanted in the operating theater following strict aseptic technique. The procedure may be carried out under spinal anesthesia, regional anesthesia [53] or local anesthesia combined with sedation. In certain cases, the option of conducting the procedure under general anesthesia will be chosen.

The patient should be monitored in the standard manner (pulse oximetry, capnography, continuous ECG and noninvasive blood pressure). It is advisable to administer an antibiotic for prophylactic purposes about 30 min before the procedure begins. Our protocol recommends the use of cefazolin 2 g intravenously.

11.1 Catheter and Needle Placement

Placement of the catheter is the aspect of the system that enables drugs to be delivered directly into the intrathecal space, so it is very important for this part of the procedure to be done properly, in order to provide a satisfactory long-term outcome.

The patient should be positioned on the operating table in such a way as to provide optimal conditions for successful placement of the needle and catheter. The most common position is lateral decubitus (preferably on the right) with the legs bent up over the trunk. The patient's two shoulders should be aligned, to prevent torsion of the trunk [54].

Fluoroscopic imaging is used to assess the bony anatomy and determine the level at which to place the needle. A paramedian approach is usually preferred, so the vertebral pedicle of the segment below the one selected for spinal puncture is marked as the entry point.

After this, the patient's skin is sterilized with antiseptic solution (preferably chlorhexidine), including both the intended area of approach and the surrounding tissues (Fig. 1).

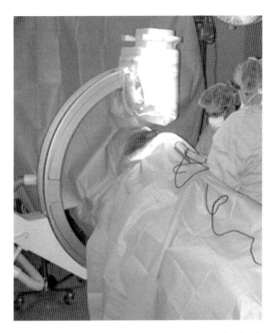

Fig. 1 Patient placed in right lateral decubitus with sterile field prepared and image intensifier positioned laterally

In some cases, the patient position achieved is not optimal because the patient is in too much pain to be placed in the lateral decubitus position, or because of rigidity or deformities in cases of spasticity. In these cases, the position can be modified, bearing in mind that if the subcutaneous pocket is to be located in the abdominal wall, the patient will have to be repositioned after insertion of the catheter.

A paramedian approach is recommended in order to protect the catheter against potential wear and tear from the spinous processes.

The needle is positioned to pass under the lamina and enter the intrathecal space at an ideal angle of 30°. In some cases, the angle may increase to 60° because of anatomical constraints, but larger angles can put excessive strain on the catheter, resulting in infusion failures [54].

Under radiological control, the needle is advanced to the subarachnoid space. Having reached this point, unobstructed flow of CSF must be observed (Fig. 2). The needle stylet is removed and the catheter is advanced, monitoring the position of its tip by fluoroscopy. Ideally, the catheter is inserted into the intrathecal space in a cephalad direction and advanced to the desired vertebral level. Unless metameric analgesia is required, it is recommended that the tip be left below the conus medullaris to prevent neurologic complications, e.g., derived from the onset of granulomas. For the same reason, it is recommended that the catheter be left in the

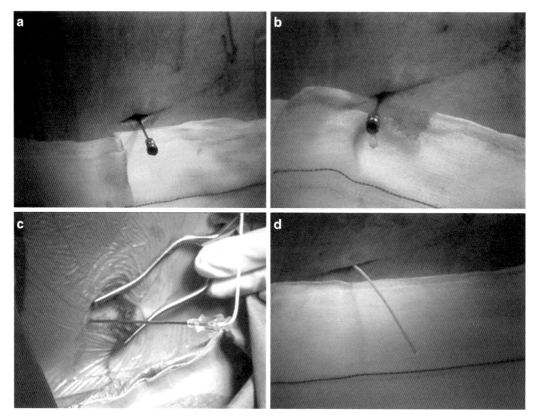

Fig. 2 (**a**) Needle inserted. (**b**) Stylet removed, showing unobstructed emission of cerebrospinal fluid. (**c**) Catheter introduced through inserted needle. (**d**) Needle removed, and catheter showing unobstructed emission of cerebrospinal fluid

dorsal space, because if located ventrally onset of a granuloma would cause motor compromise in the patient.

The lumbar incision that will subsequently contain the fixation component and the catheter is now made, freeing the needle and catheter from the surrounding tissues. Although this practice is very common among many implant physicians, at our center we make the cut and dissect the subcutaneous cellular tissue to the fascia before introducing the needle.

The greatest risk in catheter placement is damaging the nerve roots or the spinal cord. Injuries can range from transient inflammation of a nerve root to spinal lesions causing paraplegia. In order to prevent this, attention must be paid to proper patient position and correct alignment. This will assist the insertion procedure and advancement of the needle. The fluoroscopic image should be adjusted to correct for patient rotation, vertebral kyphosis or scoliosis, or other abnormalities. By using sedation with direct patient communication, the implanter can be alerted if close to the nerve

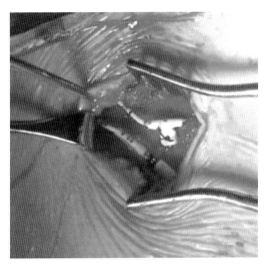

Fig. 3 Long tubular anchor anchored on the fascia and fixed with nonresorbable suture. Image shows also connection between subarachnoid catheter and extension line connecting with the pump

or spinal cord. Needle entry below the level of the conus medullaris also reduces the risk of cord damage, although this can also be caused by the advancing catheter itself [54].

Placing the needle at a proper angle and paying attention to the flow of CSF will reduce the need to make several entries into the intrathecal space, thereby lessening the risk of CSF leaks and/or PDPH occurring.

11.2 Catheter Fixation

Catheter migration is a major problem because it can lead to loss of efficacy of the system, interrupt medication delivery, and even require surgical revision. There are various different ways of securing the catheter. Options include suturing the catheter directly to the fascia, securing the catheter with the tissue surrounding it, or inserting various types of silastic plastic anchors that prevent the catheter slipping while simultaneously fixing it to the fascia [55]. These anchors should be positioned as close as possible to where the catheter enters the ligament or fascia, to prevent the catheter migrating through the gap between the anchor and the fascia. It is important to remember that all fatty tissue must be dissected as far as the fascia, as fixing the catheter elsewhere can lead to poor anchorage and subsequent catheter migration.

Although, as already mentioned, different types of anchors exist, tubular anchors are the most commonly used. These are placed over the catheter at the distal tip and slid along the catheter to its site of entry into the fascia, to abut the fascia or ligament (Fig. 3). These anchors are available mounted on an applicator to assist the sliding process and their placement on the fascia or ligament (Fig. 4).

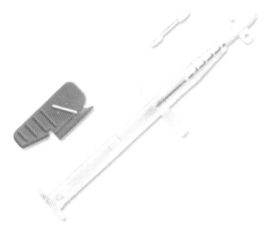

Fig. 4 Tubular anchor applicator system for spinal catheter fixation

This type of anchor may have one, two or three suture holes. When fixing them, it is important to ensure that the sutures inserted do not place strain on the catheter materials, which might lead to catheter kinking or damage [56].

The suture used to anchor the catheter must be nonabsorbable. Although many older articles recommend the use of silk sutures for fixation, this type of thread can degrade and break over time, with loss of anchorage and increased risk of catheter migration [57]. We recommend the use of coated polyester sutures (Ticron or Ethibond) as these provide durable fixation, thereby reducing the risk of migration in the long term, although to date there are no long-term studies comparing different kinds of anchor or the use of different suture types.

Catheter anchorage is an important part of the procedure, especially when the long-term stability of the system can depend on it. It is very important for the catheter to exit the spine in a smooth transition, with no bends or kinks, as with entering and leaving the anchor components. Careful securing of the catheter can improve outcomes and patient satisfaction and reduce the need for subsequent surgical revisions.

11.3 Catheter Tunneling

The chosen tunneling route should ideally be marked before the start of the surgical procedure and should take account of the anatomical peculiarities of each patient, thereby preventing the catheter from running close to or over inadvisable areas, such as the bones of the pelvis or ribs. This will also provide guidance for tissue infiltration with local anesthetic prior to the tunneling maneuver. Our group recommends adding epinephrine to the chosen local anesthetic, in order to minimize inadvertent bleeding due to tunneling.

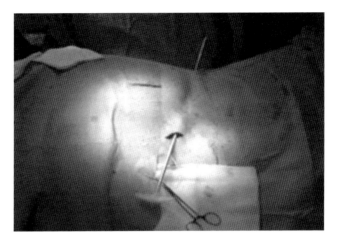

Fig. 5 Spinal catheter tunneling method

Conventional tunneling rods are of fairly large diameter. In addition, they need to follow a curved path, over long distances in most cases (from the back to the abdomen). This path may lead the tunneling rod into deep structures. In order to prevent this, the rods are made of a bendable material to adjust to the contours of the path, but we nevertheless recommend two-stage tunneling (Fig. 5). In this method, the implanter first tunnels to an intermediate point where, with the aid of an incision, the catheter is brought out. Tunneling to the subcutaneous pocket is then finished in a second step, thus completing passage of the catheter from the back to the abdominal subcutaneous pocket that will hold the pump [58].

During the course of tunneling, the implanter should check by palpation that the tunneling rod is at the proper depth, ideally in the subcutaneous cellular tissue. It should be deep enough to avoid invading the dermis, but superficial enough to prevent penetration of hazardous structures (e.g., the abdominal or thoracic cavity).

It must also be remembered that there is a risk of damaging the spinal catheter with the tunneling rod when the latter approaches the entry site of the former into the vertebral column.

Once tunneling is complete, proper CSF flow must be confirmed and the catheter should be left ready for subsequent connection to the infusion pump. In the spinal incision, it is advisable to make a loop with the excess catheter to reduce strain and help prevent catheter migration [58].

11.4 Subcutaneous Pocket Creation

Skin incision, proper dissection and separation of tissue planes, and careful hemostasis of the wound and subcutaneous pocket created are factors to focus on when creating the pump pocket [59].

One of the most important points is determining where on the anterior abdominal wall the subcutaneous pocket will be made.

Certain anatomical structures must be borne in mind when deciding on this site. The most important are the lower margin of the last rib, the iliac crest and anterior superior iliac spine, and lastly the abdominal midline and umbilicus. Ideally, about 5 cm or more should be left between the pump site and the bony margins of the ribs and bones of the pelvis. This will allow enough space when standing, sitting and lying down, avoiding bone irritation due to contact with the pump. Existing scars, skin defects and apparently infected areas of skin should also be avoided as far as possible [59]. In our experience, most patients are eligible to have the system fitted subcutaneously, although in a small percentage of them (e.g., extremely thin, emaciated patients, or even children) a subcutaneous location may be unsuitable and placement should be subfascial.

Before the incision is made, the patient should be properly anesthetized, with either local anesthesia, intravenous sedation, or both. The incision is made with the aid of a #13 or #15 scalpel, holding the skin taut with the fingers. The incision may be made to the desired depth with the scalpel, or only the dermis may be cut and the desired depth then achieved with the aid of an electric surgical knife. Normal pocket depth ranges from 1 to 3 cm; greater depths may subsequently hinder system telemetry. Once the desired depth has been reached the tissues can start to be dissected to create the pocket. The pocket can be made with surgical scissors, with the aid of the electric surgical knife, or by blunt dissection with the fingers or the blunt tip of a pair of scissors. The pocket should be about 110–120 % of total pump volume. If the pocket is too large it may allow the pump to move around inside, encourage seroma formation, and in extreme cases the pump may turn over. If the pocket is too small, it may lead to tissue pressure on the device, causing discomfort, or, in the worst-case scenario, tissue erosion with infection of the device [60].

Tissue dissection should be performed carefully. Excessive or careless manipulation can cause the subsequent appearance of seromas in the pocket, although this is not always avoidable. Likewise, careful hemostasis of the subcutaneous pocket should be effected before the device is inserted.

Once the pocket has been dissected to the proper size, the pump is filled with the medication. Having done so, the catheter previously tunneled to the subcutaneous pocket is attached to the pump. Before inserting the pump in the pocket, four stitches of nonabsorbable suture (preferably coated polyester) are placed in the bottom of the pocket so that they can then be threaded through rings located on the edge of the infusion pump, to anchor it in the proper position inside the pocket and minimize the risk of it moving or turning over (Fig. 6).

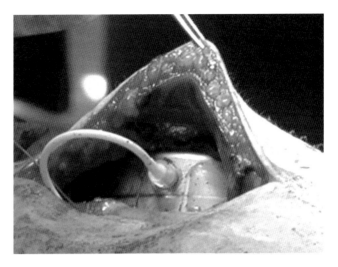

Fig. 6 Pump in the subcutaneous pocket

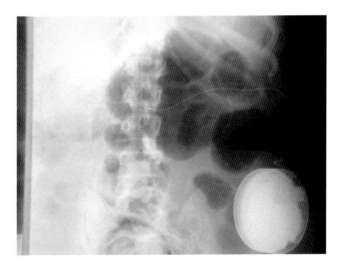

Fig. 7 Plain AP Rx for overview of implanted pump and intrathecal catheter

With the pocket made and the pump housed and fixed in its final position, the pocket is then closed. This is done in at least two layers. The choice of suture is up to the implanting physician. We recommend the use of synthetic absorbable polyglycolic acid suture (Dexon). Lastly, the skin can be closed with surgical staples. Although some authors recommend the subsequent use of compression bandages or binders to reduce the incidence of hematomas and seromas [60], in our protocol we prefer not to subject the pocket to compression if care has been exercised in dissection and hemostasis. Figure 7 shows an easy way to review the whole implanted system by using a plain X-ray film to check position and connections of all parts of the system.

References

1. Deer TR, Levy R, Prager J et al (2012) Polyanalgesic consensus conference—2012: recommendations to reduce morbidity and mortality in intrathecal drug delivery in the treatment of chronic pain. Neuromodulation 15(5):467–482

2. Lawson EF, Wallace MS (2012) Advances in intrathecal drug delivery. Curr Opin Anaesthesiol 25(5):572–576

3. Lin CP, Lin WY, Lin FS, Lee YS, Jeng CS, Sun WZ (2012) Efficacy of intrathecal drug delivery system for refractory cancer pain patients: a single tertiary medical center experience. J Formos Med Assoc 111(5):253–257

4. Upadhyay SP, Mallick PN (2012) Intrathecal drug delivery system (IDDS) for cancer pain management: a review and updates. Am J Hosp Palliat Care 29(5):388–398

5. Deer TR, Prager J, Levy R et al (2012) Polyanalgesic consensus conference 2012: recommendations for the management of pain by intrathecal (intraspinal) drug delivery: report of an interdisciplinary expert panel. Neuromodulation 15(5):436–464

6. Hamza M, Doleys D, Wells M, Weisbein J, Hoff J, Martin M, Soteropoulos C, Barreto J, Deschner S, Ketchum J (2012) Prospective study of 3-year follow-up of low-dose intrathecal opioids in the management of chronic non-malignant pain. Pain Med 13(10):1304–1313

7. Duarte RV, Raphael JH, Sparkes E, Southall JL, LeMarchand K, Ashford RL (2012) Long-term intrathecal drug administration for chronic nonmalignant pain. J Neurosurg Anesthesiol 24(1):63–70

8. Hayes C, Jordan MS, Hodson FJ, Ritchard L (2012) Ceasing intrathecal therapy in chronic non-cancer pain: an invitation to shift from biomedical focus to active management. PLoS One 7(11):e49124

9. Jackson TP, Lonergan DF, Todd RD, Martin PR (2013) Intentional intrathecal opioid detoxification in 3 patients: characterization of the intrathecal opioid withdrawal syndrome. Pain Pract 13(4):297–309

10. Deer T (2011) Selection and indications for intrathecal pump placement. In: Deer TR (ed) Atlas of implantable therapies for pain management. Springer Science, New York

11. De Andres J, Asensio-Samper JM, Fabregat-Cid G (2013) Advances in intrathecal drug delivery. Curr Opin Anaesthesiol 26:594–599

12. Deer T (2012) Polyanalgesic consensus conference 2012. Neuromodulation 15:418–419

13. Gogia V, Chaudhary P, Ahmed A, Khurana D, Mishra S, Bhatnagar S (2012) Intrathecal morphine pump for neuropathic cancer pain: a case report. Am J Hosp Palliat Care 29(5):409–411

14. Georges P, Lavand'homme P (2012) Intrathecal hydromorphone instead of the old intrathecal morphine: the best is the enemy of the good? Eur J Anaesthesiol 29(1):3–4

15. Yaksh TL, de Kater A, Dean R, Best BM, Miljanich GP (2012) Pharmacokinetic Analysis of Ziconotide (SNX-111), an Intrathecal N-Type Calcium Channel Blocking Analgesic, Delivered by Bolus and Infusion in the Dog. Neuromodulation 15(6):508–519

16. Deer TR, Prager J, Levy R et al (2012) Polyanalgesic consensus conference 2012: recommendations on trialing for intrathecal (intraspinal) drug delivery: report of an interdisciplinary expert panel. Neuromodulation 15:420–435

17. Alicino I, Giglio M, Manca F, Bruno F, Puntillo F (2012) Intrathecal combination of ziconotide and morphine for refractory cancer pain: a rapidly acting and effective choice. Pain 153(1):245–249

18. Dupoiron D, Bore F, Lefebvre-Kuntz D et al (2012) Ziconotide adverse events in patients with cancer pain: a multicenter observational study of a slow titration, multidrug protocol. Pain Physician 15(5):395–403

19. Natale M, Mirone G, Rotondo M, Moraci A (2012) Intrathecal baclofen therapy for severe spasticity: analysis on a series of 112 consecutive patients and future prospectives. Clin Neurol Neurosurg 114(4):321–325

20. Uchiyama T, Nakanishi K, Fukawa N, Yoshioka H, Murakami S, Nakano N, Kato A (2012) Neuromodulation using intrathecal baclofen therapy for spasticity and dystonia. Neurol Med Chir (Tokyo) 52(7):463–469

21. Bonouvrié LA, van Schie PE, Becher JG, van Ouwerkerk WJ, Vermeulen RJ (2012) Intrathecal baclofen for progressive neurological disease in childhood: a systematic review of literature. Eur J Paediatr Neurol 16(3):279–284

22. Bahl A, Tripathi C, McMullan J, Goddard J (2013) Novel use of intrathecal baclofen drug delivery system for periodic focal dystonia in a teenager. Neuromodulation 16(3):273–275

23. Turner M, Nguyen HS, Cohen-Gadol AA (2012) Intraventricular baclofen as an alternative to intrathecal baclofen for intractable spasticity or dystonia: outcomes and technical considerations. J Neurosurg Pediatr 10(4):315–319

24. Chiodo AE, Saval A (2012) Intrathecal baclofen for the treatment of spinal myoclonus: a case series. J Spinal Cord Med 35(1):64–67

25. Hoarau X, Richer E, Dehail P, Cuny E (2012) A 10-year follow-up study of patients with severe traumatic brain injury and dysautonomia treated with intrathecal baclofen therapy. Brain Inj 26(7–8):927–940

26. Van der Plas AA, van Rijn MA, Marinus J, Putter H, van Hilten JJ (2013) Efficacy of intrathecal baclofen on different pain qualities in complex regional pain syndrome. Anesth Analg 116(1):211–215

27. Pritula SL, Fox MA, Ayyangar R (2012) Weight changes in children receiving intrathecal baclofen for the treatment of spasticity. J Pediatr Rehabil Med 5(3):197–201

28. Bensmail D, Marquer A, Roche N, Godard AL, Lofaso F, Quera-Salva MA (2012) Pilot study assessing the impact of intrathecal baclofen administration mode on sleep-related respiratory parameters. Arch Phys Med Rehabil 93(1):96–99

29. Murakami M, Hirata Y, Kuratsu J (2012) Paradoxical worsening of spasticity and pain in the lower extremities after increasing the dose of intrathecal baclofen—case report. Neuromodulation 15(1):39–40

30. Dickey M, Rice M, Kinnett D, Lambert R, Donaver S, Gerber M, Staat M (2013) Infectious complications of intrathecal baclofen pump devices in a pediatric population. Pediatr Infect Dis J 32(7):715–722

31. Monsalve V (2003) Neuromodulación. Valoración psicológica y estrategias de aplicación. Rev Soc Esp Dolor 6(5):357–362

32. Gybels J, Erdine S, Maeyaert J (1998) Neuromodulation of pain. A consensus statement prepared in Brussels 16–18 January 1998 by the following task force of the European Federation of IASP Chapters (EFIC). Eur J Pain 2(3):203–209

33. Monsalve V, de Andres JA, Valia JC (2000) Application of a psychological decision algorithm for the selection of patients susceptible to implantation of neuromodulation systems for the treatment of chronic pain. A proposal. Neuromodulation 3:191–200

34. Avila A, Jimenez F (2002) Adaptación española del Inventario Multifásico de Personalidad de Minnesota-2 (MMPI-2). TEA ediciones, Madrid

35. Avila A, Jimenez F (2004) Adaptación española del Inventario Clínico Multiaxial de Millon-II (MCMI-II). TEA Ediciones, Madrid

36. Soriano J, Monsalve V (2004) Validación del Cuestionario de Afrontamiento al Dolor Crónico reducido (CAD-R). Rev Soc Esp Dolor 11(7):407–414

37. Rodríguez L, Cano FJ, Blanco A (2004) Evaluación de las estrategias de afrontamiento del dolor crónico. Actas Esp Psiquiatr 32(2):82–91

38. Tejero A, Guimera EM, Farré JM (1986) Uso clínico del HAD (hospital anxiety and depression scale) en población psiquiátrica: un estudio de su sensibilidad, fiabilidad y validez. Rev Dpto Psiquiatria de la F De Medicina de Barcelona 13:233–238

39. Spielberger CD, Gorsuch RL, Lushene RE (1982) Inventario de Ansiedad Estado-Rasgo (STAI), 1968. TEA Ediciones, Madrid, Adaptación en castellano

40. Vázquez C, Sanz J (1997) Fiabilidad y valores normativos de la versión española del Inventario para la Depresión de Beck de 1978. Clínica y Salud 8:403–422

41. Williams DA, Thorn BE (1989) An empirical assessment of pain beliefs. Pain 36(3):351–358

42. Huskisson EC (1983) Visual analogue scales. In: Melzack R (ed) Pain measurement and assessment. Raven, New York, pp 33–37

43. Lázaro C, Bosch F, Torrubia R, Baños JE (1994) The development of a Spanish questionnaire for assessing pain: preliminary data concerning reliability and validity. Eur J Psychol Assess 10(2):145–151

44. Monsalve V, Soriano J, De Andrés J (2006) Utilidad del Índice de Lattinen en la evaluación del dolor crónico: relaciones con afrontamiento y calidad de vida. Rev Soc Esp Dolor 13(4):216–230

45. González Andrés VL (1999) Evaluación del uso de las bombas de infusión programables para la administración de fármacos en el espacio intratecal. ACSA, Sevilla, Spain

46. López Millán J (2006) Análisis Coste-Efectividad de la Terapia Intratecal. Guía de Infusión Espinal en Espasticidad. Grupo Español de Neuroestimulación

47. Smith TJ et al (2005) An implantable drug delivery system (IDDS) for refractory cancer pain provides sustained pain control, less drug-related toxicity and possibly better survival compared with comprehensive medical management. Ann Oncol 16:825–833

48. Deer TR, Prager J, Levy R, Rathmell J, Buchser E, Burton A, Caraway D, Cousins M, De Andrés J, Diwan S, Erdek M, Grigsby E, Huntoon M, Jacobs MS, Kim P, Kumar K, Leong M, Liem L, McDowell GC 2nd, Panchal S, Rauck R, Saulino M, Sitzman BT, Staats P, Stanton-Hicks M, Stearns L, Wallace M, Willis KD, Witt W, Yaksh T, Mekhail N (2012) Polyanalgesic Consensus Conference--2012: consensus on diagnosis, detection, and treatment of catheter-tip granulomas (inflammatory masses). Neuromodulation 15(5):483–495

49. De Andrés J, Tatay Vivò J, Palmisani S, Villanueva Pérez VL, Mínguez A (2010) Intrathecal granuloma formation in a patient receiving long-term spinal infusion of tramadol. Pain Med 11(7):1059–1062

50. De Andrés J, Palmisani S, Villanueva Pérez VL, Asensio J, López-Alarcón MD (2010) Can an intrathecal, catheter-tip-associated inflammatory mass reoccur? Clin J Pain 26(7):631–634

51. Hayek SM, Deer TR, Pope JE, Panchal SJ, Patel VB (2011) Intrathecal therapy for cancer and non-cancer pain. Pain Physician 14: 219–248

52. Krames E (2002) Implantable devices for pain control: spinal cord stimulation and intrathecal therapies. Best Pract Res Clin Anaesthesiol 16:619–649

53. Asensio-Samper JM, De Andrés-Ibáñez J, Fabregat-Cid G, Villanueva Pérez V, Alarcón L (2010) Ultrasound-guided transversus abdominis plane block for spinal infusion and neurostimulation implantation in two patients with chronic pain. Pain Pract 10(2):158–162

54. Deer T (2011) Placement of intrathecal needle and catheter for chronic infusion. In: Deer TR (ed) Atlas of implantable therapies for pain management. Springer Science, New York

55. Deer T (2011) Securing and anchoring permanent intrathecal catheters. In: Deer TR (ed) Atlas of implantable therapies for pain management. Springer Science, New York

56. Follett K, Burchiel K, Deer T et al (2003) Prevention of intrathecal drug delivery catheter-related complications. Neuromodulation 6(1): 3–41

57. McIntyre P, Deer T, Hayek S (2007) Complications of spinal infusion therapies. Tech Reg Anesth Pain Manag 11(3):183–192

58. Deer T (2011) Tunneling permanent intrathecal catheters. In: Deer TR (ed) Atlas of implantable therapies for pain management. Springer Science, New York

59. Deer T (2011) Pocketing for intrathecal drug delivery systems. In: Deer TR (ed) Atlas of implantable therapies for pain management. Springer Science, New York

60. Turner J, Sears J, Loeser D (2007) Programmable intrathecal opioid delivery systems for chronic noncancer pain: a systemic review of effectiveness and complications. Clin J Pain 23(2):180–195

INDEX

Kewal K. Jain (ed.), *Drug Delivery System*, Methods in Molecular Biology, vol. 1141,
DOI 10.1007/978-1-4939-0363-4, © Springer Science+Business Media New York 2014

Printed by Printforce, the Netherlands